Arduino by Example

Design and build fantastic projects and devices using the Arduino platform

Adith Jagadish Boloor

[PACKT] open source *
PUBLISHING   community experience distilled

BIRMINGHAM - MUMBAI

Arduino by Example

First published: September 2015

Production reference: 1090915

Published by Packt Publishing Ltd.
Livery Place
35 Livery Street
Birmingham B3 2PB, UK.

ISBN 978-1-78528-908-8

www.packtpub.com

Credits

Author
Adith Jagadish Boloor

Reviewers
Tim Gorbunov

Francis Perea

Commissioning Editor
Edward Bowkett

Acquisition Editor
Vivek Anantharaman

Content Development Editor
Arwa Manasawala

Technical Editor
Mitali Somaiya

Copy Editors
Trishya Hajare

Kevin McGowan

Project Coordinator
Shweta H Birwatkar

Proofreader
Safis Editing

Indexer
Priya Sane

Production Coordinator
Nitesh Thakur

Cover Work
Nitesh Thakur

About the Author

Adith Jagadish Boloor was born in Mangalore, India. He grew up tinkering with toys and gadgets that kindled his interest in how things work. His admiration for science and technology, specifically in the fields of robotics, 3D printing, and smart systems, grew into a passion that he is working towards, nurturing it into a career. He completed his higher studies at Purdue University, USA and Shanghai Jiao Tong University, China and is working towards obtaining a masters degree in robotics.

Adith has experience working on robots ranging from simple obstacle—avoiding robots built at home to complex humanoid robots such as the Darwin-OP in Purdue University's research lab. He has coauthored a research paper and has two patents on his name.

He enjoys traveling and grabs every opportunity he can to explore the different parts of the world. He is also an international chess player.

Acknowledgements

I would like to thank my dad, Jagadish, mom, Bharathi, and sister, Anvitha, for their unconditional support through the duration of writing this book. I would also like to give a lot of credit to Arwa Manasawala, Vivek Anantharaman, and the entire team at Packt Publishing for putting up with me, guiding me, and most of all, giving me the wonderful opportunity to share what I have learned over the years with those looking for it.

This book couldn't have been written if I myself didn't have the knowledge and experience about the subject. I owe this to my mentors. I would like to thank Frits Lyneborg, the creator of letsmakerobots.com, a website that ignited my passion for robotics. I am indebted to Dr Eric Matson and the incredible team at Purdue's M2M research lab, who shared my curiosity in technology and helped me become a better roboticist.

Last but not the least, I would like to thank each and every friend and colleague at Purdue University, without whom this book wouldn't be nearly as good as I hoped.

About the Reviewers

Tim Gorbunov was born in the USA. At a young age, he fell in love with building and constructing things, just like his dad. Tim became very good at origami and started to sell it at elementary school. As he grew up, Tim leaned more towards electronics because it fascinated him more than any other hobby. Creating circuits that buzzed or flashed was one of Tim's favorite things to do. As time went by, he started exploring more advanced electronics and programming, and from that point on, he became more and more knowledgeable about electronics. He got hired to help create cymatic light shows at CymaSpace. At this company that specializes in sound-reactive technologies, he helped start Audiolux devices by helping them design their products. Tim has many other hobbies, but he does a good job at implementing his electronic ideas in his activities. One example of this is a fishing boat motor controller that allows the user to throttle and shift using a joystick, which is all based on the Arduino in his custom printed circuit board.

I would like to thank books, such as this one which I was privileged to review, and the Internet for allowing me to learn so many cool things about the Arduino and the electronics world.

Francis Perea is a professional education professor at Consejería de Educación de la Junta de Andalucía in Spain with more than 14 years of experience. He specializes in system administration, web development, and content management systems. In his spare time, he works as a freelancer and collaborates, among others, with ñ multimedia, a small design studio in Córdoba, working as a system administrator and main web developer.

He has also collaborated as a technical reviewer on *SketchUp 2013 for Architectural Visualization*, *Arduino Home Automation*, *Internet of Things with the Arduino Yún*, and *Arduino Cookbook* by Packt Publishing.

When not sitting in front of a computer or tinkering in his workshop, he can be found mountain biking or kite surfing or, as a beekeeper, taking care of his hives in Axarquía County, where he lives.

I would like to thank my wife, Salomé, and our three kids, Paula, Álvaro, and Javi, for all the support they gave me, even when we all were busy. There are no words to express my gratitude towards them.

I would also like to thank my colleagues at ñ multimedia and my patient students. The need to be at the level you demand is what keeps me going forward.

www.PacktPub.com

Support files, eBooks, discount offers, and more

For support files and downloads related to your book, please visit www.PacktPub.com.

Did you know that Packt offers eBook versions of every book published, with PDF and ePub files available? You can upgrade to the eBook version at www.PacktPub. com and as a print book customer, you are entitled to a discount on the eBook copy. Get in touch with us at service@packtpub.com for more details.

At www.PacktPub.com, you can also read a collection of free technical articles, sign up for a range of free newsletters and receive exclusive discounts and offers on Packt books and eBooks.

https://www2.packtpub.com/books/subscription/packtlib

Do you need instant solutions to your IT questions? PacktLib is Packt's online digital book library. Here, you can search, access, and read Packt's entire library of books.

Why subscribe?

- Fully searchable across every book published by Packt
- Copy and paste, print, and bookmark content
- On demand and accessible via a web browser

Free access for Packt account holders

If you have an account with Packt at www.PacktPub.com, you can use this to access PacktLib today and view 9 entirely free books. Simply use your login credentials for immediate access.

Table of Contents

Preface

With the growing interest in home-made, weekend projects among students and hobbyists alike, Arduino offers an innovative and feasible platform to create projects that promote creativity and technological tinkering. Whether you are an experienced programmer or a person who wants to enter the world of electronics and do not know how to begin, this book will teach you the necessary skills that you will need to successfully build Arduino-powered projects that have real-life implications. Initially, you will learn how to get started with the Arduino platform. The example-based, project-oriented setup of this book will progressively grow in complexity to expand your knowledge. With what you will learn, you will be able to construct your own devices.

What this book covers

Chapter 1, *Getting Started with Arduino*, introduces the reader to the Arduino platform, beginning with acquiring the necessary components and installing the software to write your first program and see the magic begin.

Chapter 2, *Digital Ruler*, brings in commonly used Arduino-friendly components such as an ultrasound sensor and a small programmable LCD panel, and puts them together to create a digital ruler, which is capable of measuring distances using the sensor and displaying them in real time on the LCD screen.

Chapter 3, *Converting Finger Gestures to Text*, makes use of a relatively new line of sensors such as a fully functional touch sensor. The basic algorithms are taught that allow the Arduino to translate finger gestures into corresponding characters that are then displayed graphically using a commonly used software called Processing.

Chapter 4, Burglar Alarm – Part 1, introduces the reader to using PIR sensors or motion sensors, implementing a remote camera with Arduino, and linking the Arduino to a smart phone. Additionally, the reader will learn about Python and how it interfaces with Arduino.

Chapter 5, Burglar Alarm – Part 2, combines the elements learned in the preceding project with a project that uses a sensor to detect motion at an entry point, which triggers a security camera to take the intruder's photo via Bluetooth and sends that image to your smart phone.

Chapter 6, Home Automation – Part 1, follows the sophisticated security system's path. This chapter involves connecting the Arduino to the Wi-Fi network using an electro-magnetic switch called a relay to control an electric appliance and communicating to it using Telnet.

Chapter 7, Home Automation – Part 2, uses the Arduino to create a simple home automation system operating within the bounds of the Wi-Fi that would allow the user to control an appliance using a computer, smart phone, and their voice.

Chapter 8, Robot Dog – Part 1, revolves around building a four-legged robot dog from scratch. This part teaches you about the Arduino MEGA board, servos, and stand-alone power requirements for the board.

Chapter 9, Robot Dog – Part 2, involves using household items to build the chassis of the dog and then completing the circuit using the Arduino MEGA board and a lot of servos. This is where the bulk of the actual construction of the robot dog lies.

Chapter 10, Robot Dog – Part 3, acts as the icing on the cake. The reader will finally finish building the robot and will learn to calibrate and teach (program) the robot to stand, walk, and play. Also, finally, speech recognition will be implemented so that the dog can actually listen to the user.

What you need for this book

The primary software required are as follows:

- Arduino IDE
- Processing IDE
- Python 2.7
- BitVoicer
- Teraterm
- Putty

Who this book is for

Arduino by Example is intended for anyone interested in, or keen to get into, the world of electronics, robotics, Internet of Things, and security systems. The reader will learn to build projects involving touch sensors, home automation, robots, and home security. Even experienced Arduino veterans can pick this book up and get a lot out of it. Programming knowledge is not required for using this book. This book teaches the reader the basics and will quickly and progressively guide them through more complex topics.

Conventions

In this book, you will find a number of text styles that distinguish between different kinds of information. Here are some examples of these styles and an explanation of their meaning.

Code words in text, database table names, folder names, filenames, file extensions, pathnames, dummy URLs, user input, and Twitter handles are shown as follows: "Call it `helloworld.py` and press finish."

A block of code is set as follows:

```
void loop() {
digitalWrite(led, HIGH);    // turn the LED on (HIGH is the voltage
level)
delay(1000);                // wait for a second
digitalWrite(led, LOW);     // turn the LED off by making the voltage
LOW
delay(1000);                // wait for a second
}
```

When we wish to draw your attention to a particular part of a code block, the relevant lines or items are set in bold:

```
print("Burglar Alarm Program Initializing")
init("< your push overtoken>")
CLIENT_ID = "<your client ID>"
PATH = "C:\\<your python folder>\\mug_shot.jpg"
im = pyimgur.Imgur(CLIENT_ID)
```

Any command-line input or output is written as follows:

```
sudo apt-get update && sudo apt-get install arduino arduino-core
```

New terms and **important words** are shown in bold. Words that you see on the screen, for example, in menus or dialog boxes, appear in the text like this: "The **Board** option opens up all the different boards that the software supports."

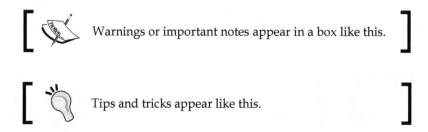

Warnings or important notes appear in a box like this.

Tips and tricks appear like this.

Reader feedback

Feedback from our readers is always welcome. Let us know what you think about this book—what you liked or disliked. Reader feedback is important for us as it helps us develop titles that you will really get the most out of.

To send us general feedback, simply e-mail feedback@packtpub.com, and mention the book's title in the subject of your message.

If there is a topic that you have expertise in and you are interested in either writing or contributing to a book, see our author guide at www.packtpub.com/authors.

Customer support

Now that you are the proud owner of a Packt book, we have a number of things to help you to get the most from your purchase.

Downloading the example code

You can download the example code files from your account at http://www.packtpub.com for all the Packt Publishing books you have purchased. If you purchased this book elsewhere, you can visit http://www.packtpub.com/support and register to have the files e-mailed directly to you.

Downloading the color images of this book

We also provide you with a PDF file that has color images of the screenshots/ diagrams used in this book. The color images will help you better understand the changes in the output. You can download this file from: `http://www.packtpub.com/sites/default/files/downloads/1234OT_ColorImages.pdf`.

Errata

Although we have taken every care to ensure the accuracy of our content, mistakes do happen. If you find a mistake in one of our books—maybe a mistake in the text or the code—we would be grateful if you could report this to us. By doing so, you can save other readers from frustration and help us improve subsequent versions of this book. If you find any errata, please report them by visiting `http://www.packtpub.com/submit-errata`, selecting your book, clicking on the **Errata Submission Form** link, and entering the details of your errata. Once your errata are verified, your submission will be accepted and the errata will be uploaded to our website or added to any list of existing errata under the Errata section of that title.

To view the previously submitted errata, go to `https://www.packtpub.com/books/content/support` and enter the name of the book in the search field. The required information will appear under the **Errata** section.

Piracy

Piracy of copyrighted material on the Internet is an ongoing problem across all media. At Packt, we take the protection of our copyright and licenses very seriously. If you come across any illegal copies of our works in any form on the Internet, please provide us with the location address or website name immediately so that we can pursue a remedy.

Please contact us at `copyright@packtpub.com` with a link to the suspected pirated material.

We appreciate your help in protecting our authors and our ability to bring you valuable content.

Questions

If you have a problem with any aspect of this book, you can contact us at `questions@packtpub.com`, and we will do our best to address the problem.

1
Getting Started with Arduino

Hello there! If you are reading this book right now, it means that you've taken your first step to make fascinating projects using Arduinos. This chapter will teach you how to set up an Arduino and write your first Arduino code.

You'll be in good hands whilst you learn some of the basics aspects of coding using the Arduino platform; this will allow you to build almost anything including robots, home automation systems, touch interfaces, sensory systems, and so on. Firstly, you will learn how to install the powerful Arduino software, then set that up, followed by hooking up your Arduino board and, after making sure that everything is fine and well, you will write your first code! Once you are comfortable with that, we will modify the code to make it do something more, which is often what Arduino coders do. We do not just create completely new programs; often we build on what has been done before, to make it better and more suited to our objectives. The contents of this chapter are divided into the following topics:

- Prerequisites
- Setting up
- Hello World
- Summary

Prerequisites

Well, you can't jump onto a horse without putting on a saddle first, can you? This section will cover what components you need to start coding on an Arduino. These can be purchased from your favorite electrical hobby store or simply ordered online.

Materials needed

- 1x Arduino-compatible board such as an Arduino UNO
- 1x USB cable A to B
- 2x LEDs
- 2x 330Ω resistors
- A mini breadboard
- 5x male-to-male jumper wires

Note

The UNO can be substituted for any other Arduino board (Mega, Leonardo, and so on) for most of the projects. These boards have their own extra features. For example, the Mega has almost double the number of I/O (input/output) pins for added functionality. The Leonardo has a feature that enables it to control the keyboard and mouse of your computer.

Setting up

This topic involves downloading the Arduino software, installing the drivers, hooking up the Arduino, and understanding the IDE menus.

Downloading and installing the software

Arduino is open source-oriented. This means all the software is free to use non-commercially. Go to `http://arduino.cc/en/Main/Software` and download the latest version for your specific operating system. If you are using a Mac, make sure you choose the right Java version; similarly on Linux, download the 32-or 64-bit version according to your computer.

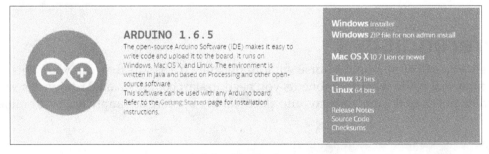

Arduino download page

Windows

Once you have downloaded the setup file, run it. If it asks for administrator privileges, allow it. Install it in its default location (C:\Program Files\Arduino or C:\Program Files (x86)\Arduino). Create a new folder in this location and rename it My Codes or something where you can conveniently store all your programs.

Mac OS X

Once the ZIP file has finished downloading, double-click to expand it. Copy the Arduino application to the Applications folder. You won't have to install additional drivers to make the Arduino work since we will be using only the Arduino UNO and MEGA throughout the book. You're all set.

If you didn't get anything to work, go to https://www.arduino.cc/en/guide/macOSX.

Linux (Ubuntu 12.04 and above)

Once you have downloaded the latest version of Arduino from the preceding link, install the compiler and the library packages using the following command:

```
sudo apt-get update && sudo apt-get install arduino arduino-core
```

If you are using a different version of Linux, this official Arduino walkthrough at http://playground.arduino.cc/Learning/Linux will help you out.

Connecting the Arduino

It is time to hook up the Arduino board. Plug in the respective USB terminals to the USB cable and the tiny LEDs on the Arduino should begin to flash.

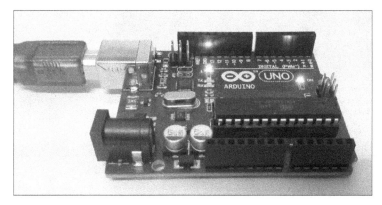

Arduino UNO plugged in

If the LEDs didn't turn on, ensure that the USB port on your computer is functioning and make sure the cable isn't faulty. If it still does not light up, there is something wrong with your board and you should get it checked.

Windows

The computer will begin to install the drivers for the Arduino by itself. If it does not succeed, do the following:

1. Open **Device Manager**.
2. Click on **Ports (COM & LPT)**.
3. Right-click on **Unknown Device** and select **Properties**.
4. Click on **Install Driver** and choose **browse files on the computer**.
5. Choose the drivers folder in the previously installed Arduino folder.

The computer should say that your Arduino UNO (USB) has been successfully installed on COM port (xx). Here xx refers to a single or double digit number. If this message didn't pop up, go back to the **Device Manager** and check if it has been installed under **COM** ports.

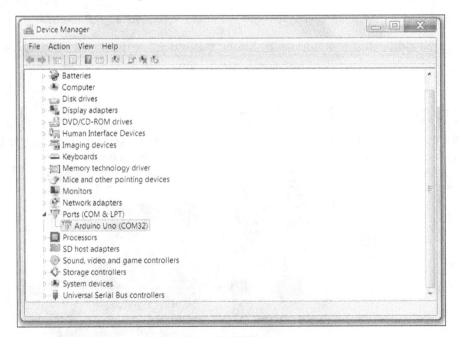

Arduino UNO COM port

Remember the (COMxx) port that the Arduino UNO was installed on.

Mac OS X

If you are using Mac OS, a dialog box will tell you that a new network interface has been detected. Click **Network Preferences** and select **Apply**. Even though the Arduino board may show up as **Not Configured**, it should be working perfectly.

Linux

You are ready to go.

The serial ports for Mac OS and Linux will be obtained once the Arduino software has been launched.

The Arduino IDE

The Arduino software, commonly referred to as the Arduino IDE (Integrated Development Environment), is something that you will become really familiar with as you progress through this book. The IDE for Windows, Mac OS, and Linux is almost identical. Now let's look at some of the highlights of this software.

Arduino IDE

This is the window that you will see when you first start up the IDE. The tick/ check mark verifies that your code's syntax is correct. The arrow pointing right is the button that uploads the code to the board and checks if the code has been changed since the last upload or verification. The magnifying glass is the **Serial Monitor**. This is used to input text or output debugging statements or sensor values.

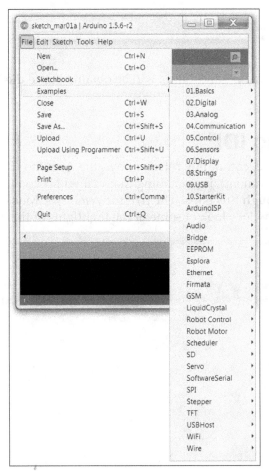

Examples of Arduino

All Arduino programmers start by using one of these examples. Even after mastering Arduino, you will still return here to find examples to use.

Arduino tools

The screenshot shows the tools that are available in the Arduino IDE. The **Board** option opens up all the different boards that the software supports.

Hello World

The easiest way to start working with Arduinos begins here. You'll learn how to output print statements. The Arduino uses a **Serial Monitor** for displaying information such as print statements, sensor data, and the like. This is a very powerful tool for debugging long codes. Now for your first code!

Writing a simple print statement

Open up the Arduino IDE and copy the following code into a new sketch:

```
void setup() {
Serial.begin(9600);
Serial.println("Hello World!");
}

void loop() {
}
```

Open **Tools | Board** and choose **Arduino UNO,** as shown in the following screenshot:

Open **Tools | Port** and choose the appropriate port (remember the previous COM xx number? select that), as shown in the following screenshot. For Mac and Linux users, once you have connected the Arduino board, going to **Tools | Serial Port** will give you a list of ports. The Arduino is typically something like /dev/tty. usbmodem12345 where *12345* will be different.

Selecting the Port

Finally, hit the Upload button. If everything is fine, the LEDs on the Arduino should start flickering as the code is uploaded to the Arduino. The code will then have uploaded to the Arduino.

To see what you have accomplished, click on the **Serial Monitor** button on the right side and switch the baud rate on the **Serial Monitor** window to 9600.

You should see your message Hello World! waiting for you there.

LED blink

That wasn't too bad but it isn't cool enough. This little section will enlighten you, literally.

Open up a new sketch.

Go to **File** | **Examples** | 01. **Basics** | **Blink**.

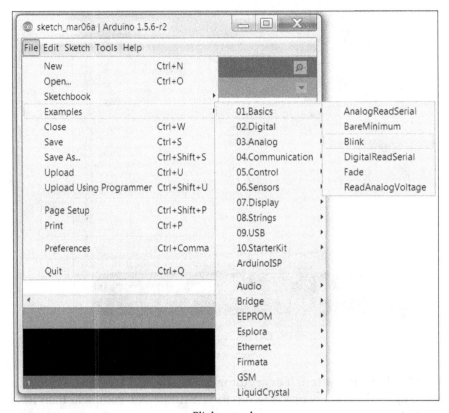

Blink example

Before we upload the code, we need to make sure of one more thing. Remember the LED that we spoke about in the prerequisites? Let's learn a bit about it before plugging it in, as shown in the following image:

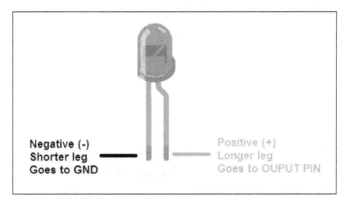

LED basics

We will make use of it now. Plug in the LED such that the longer leg goes into pin 13 and the shorter leg goes into the GND pin, as in the following:

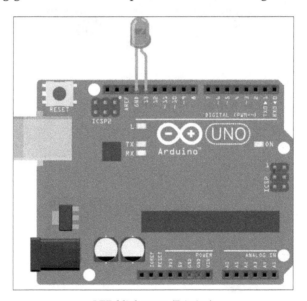

LED blink setup (Fritzing)

This diagram is made using software called Fritzing. This software will be used in future projects to make it cleaner to see and easier to understand as compared to a photograph with all the wires running around. Fritzing is open source software which you can learn more about at www.fritzing.org.

Arduino LED setup

Upload the code. Your LED will start blinking, as shown in the following image.

A lit up LED

Isn't it just fascinating? You just programmed your first hardware. There's no stopping you now. Before advancing to the next chapter, let's see what the code does and what happens when you change it.

This is the blink example code that you just used:

```
/*
Blink
Turns on an LED on for one second, then off for one second,
repeatedly.

This example code is in the public domain.
*/

//Pin 13 has an LED connected on most Arduino boards.
//give it a name:
int led = 13;

//the setup routine runs once when you press reset:
void setup() {
// initialize the digital pin as an output.
pinMode(led, OUTPUT);
}

//the loop routine runs over and over again forever:
void loop() {
  digitalWrite(led, HIGH);   // turn the LED on (HIGH is the voltage
level)
  delay(1000);               // wait for a second
  digitalWrite(led, LOW);    // turn the LED off by making the voltage
LOW
  delay(1000);               // wait for a second
}
```

We have three major sections in this code. This format will be used for most of the projects in the book.

```
int led = 13;
```

This line simply stores the numerical PIN value onto a variable called `led`.

```
void setup() {
// initialize the digital pin as an output.
pinMode(led, OUTPUT);
}
```

This is the `setup` function. Here is where you tell the Arduino what is connected on each used pin. In this case, we tell the Arduino that there is an output device (LED) on pin 13.

```
void loop() {
digitalWrite(led, HIGH);    // turn the LED on (HIGH is the voltage
level)
delay(1000);                // wait for a second
digitalWrite(led, LOW);     // turn the LED off by making the voltage
LOW
delay(1000);                // wait for a second
}
```

This is the `loop` function. It tells the Arduino to keep repeating whatever is inside it in a sequence. The `digitalWrite` command is like a switch that can be turned ON (`HIGH`) or OFF (`LOW`). The `delay(1000)` function simply makes the Arduino wait for a second before heading to the next line.

If you wanted to add another LED, you'd need some additional tools and some changes to the code. This is the setup that you want to create.

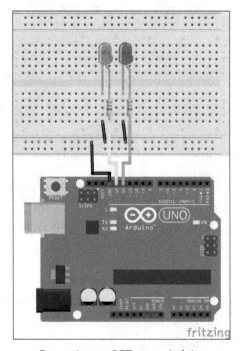

Connecting two LEDs to an Arduino

If this is your first time using a breadboard, take some time to make sure all the connections are in the right place. The colors of the wires don't matter. However, GND is denoted using a black wire and VCC/5V/PWR is denoted with a red wire. The two resistors, each connected in series (acting like a connecting wire itself) with the LEDs, limit the current flowing to the LEDs, making sure they don't blow up.

As before, create a new sketch and paste in the following code:

```
/*
Double Blink
Turns on and off two LEDs alternatively for one second each
repeatedly.

This example code is in the public domain.
*/

int led1 = 12;
int led2 = 13;

void setup() {
// initialize the digital pins as an output.
pinMode(led1, OUTPUT);
pinMode(led2, OUTPUT);

// turn off LEDs before loop begins
digitalWrite(led1, LOW);   // turn the LED off (LOW is the voltage
level)
digitalWrite(led2, LOW);   // turn the LED off (LOW is the voltage
level)
}


//the loop routine runs over and over again forever:
void loop() {
digitalWrite(led1, HIGH);   // turn the LED on (HIGH is the voltage
level)
digitalWrite(led2, LOW);   // turn the LED off (LOW is the voltage
level)
delay(1000);                // wait for a second
digitalWrite(led1, LOW);   // turn the LED off (LOW is the voltage
level)
digitalWrite(led2, HIGH);   // turn the LED on (HIGH is the voltage
level)
delay(1000);                // wait for a second
}
```

Once again, make sure the connections are made properly, especially the positive LEDs (the longer one to OUTPUT PIN) and the negative (the shorter to the GND) terminals. Save the code as `DoubleBlink.ino`. Now, if you make any changes to it, you can always retrieve the backup.

Upload the code. 3... 2... 1... And there you have it, an alternating LED blink cycle created purely with the Arduino. You can try changing the delay to see its effects.

For the sake of completeness, I would like to mention that you could take this mini-project further by using a battery to power the system and decorate your desk/room/house. More on how to power the Arduino will be covered in subsequent chapters.

Summary

You have now completed the basic introduction to the world of Arduino. In short, you have successfully set up your Arduino and have written your first code. You also learned how to modify the existing code to create something new, making it more suitable for your specific needs. This methodology will be applied repeatedly while programming, because almost all the code available is open source and it saves time and energy.

In the next chapter, we will look into sensors and displays. You will build a digital ruler that you can use to measure short distances. It will consist of an ultrasound sensor to compute distance and a small LCD screen to display it. Additionally, we will look at safely powering the Arduino board using a battery so that you are not dependent on your computer for USB power every time.

2
Digital Ruler

You've made it to chapter 2! Congrats! From now on things are going to get a bit complicated as we try to make the most of the powerful capabilities of the Arduino micro controller. In this chapter we are going to learn how to use a sensor and an LCD board to create a digital LCD ruler.

Put simply, we will use the ultrasound sensor to gauge the distance between the sensor and an object. We will use the Arduino and some math to convert the distance into meaningful data (cm, inches) and finally display this on the LCD.

- Prerequisites
- Using an ultrasound sensor
- Hooking up an LCD to the Arduino
- Displaying the sensor data on the LCD
- Summary

Prerequisites

The following is a list of materials that you'll need to start coding on an Arduino; these can be purchased from your favorite electrical hobby store or simply ordered online:

- 1 x Arduino-compatible board such as the UNO
- 1 x USB cable A to B 1 x HC−SR04 ultrasound sensor
- 1 x I2C LCD1602
- 10 x male to male wires
- 9V battery with 2.1 mm barrel jack connector (optional)
- Laser pointer (optional)

Components such as the LCD panel and the ultrasonic sensor can be found in most electronic hobby stores. If they are unavailable in a store near you, you will find online stores that ship worldwide.

A bit about the sensor

The SR04 is a very powerful and commonly used distance/proximity sensor. And that is what we are going to look at first. The SR04 sensor emits ultrasonic waves which are sound waves at such a high frequency (40 kHz) that they are inaudible to humans. When these waves come across an object, some of them get reflected. These reflected waves get picked up by the sensor and it calculates how much time it took for the wave to return. It then converts this time into distance.

We are firstly going to use this sensor to make a simple proximity switch. Basically, when you bring an object closer than the set threshold distance, an LED is going to light up.

This is the circuit that we need to construct. Again, be very careful about where everything goes and make sure there are no mistakes. It is very easy to make a mistake, no matter how much experience you've had with Arduinos.

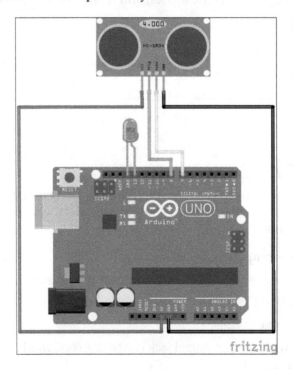

In reality it is going to look something like this, much messier than the **Fritzing** circuit depicted in the previous screenshot:

Open a new sketch on the Arduino IDE and load the SR04_Blink.ino program that came with this book.

Save the code as SR04_Blink.ino in your codes directory. This enables us to keep the supplied code as a backup if we tweak it and end up messing up the program. Do this in every instance. Now, once more, check and ensure that the pins match the topmost lines of the code. Upload the code. Now open the **Serial Monitor** on the Arduino IDE and select 9600 as the baud rate. Place your hand or a flat surface (a book) in front of it and keep changing the distance.

You should be able to see the sensor distances being displayed on the screen, as in the following screenshot:

```
COM32                                    [_] [□] [X]

                                              [ Send ]
Outside sensor range                                   ▲
41 cm
29 cm
29 cm
28 cm
28 cm                                                  ▤
27 cm
24 cm
21 cm
21 cm
21 cm
21 cm
21 cm
19 cm
17 cm
16 cm
16 cm
16 cm
17 cm
18 cm
19 cm
17 cm
18 cm
18 cm
18 cm                                                  ▼
☐ Autoscroll          No line ending ▼   9600 baud ▼
```

It says `Outside sensor range` if the sensor is picking up values greater than 200 cm because that is the most it can measure. Otherwise, if you make it point at nothing at a distance, it will still display around 200 cm because that is its range.

You will notice that, as you bring the object closer to the sensor than 15cm, it lights up. This is because the threshold is set at 15cms, as you can see in the following code snippet:

```
    if (distance < 15) {  // Threshold set to 15 cm; LED turns off if
object distance < 15cms
      digitalWrite(ledPin,HIGH);
}
  else {
      digitalWrite(ledPin,LOW);
  }
  if (distance >= 200 || distance <= 0){
      Serial.println("Outside sensor range");
  }
  else {
      Serial.print(distance); // prints the distance on the serial
monitor
      Serial.println(" cm");
  }
  delay(500); // wait time between each reading
}
```

This is the same principle used in cars (the beeping sound while reversing if you are too close to a wall), except usually it is an infrared sensor that emits light instead of sound.

In the code, we had the following line:

```
distance = duration / 58
```

This line is used to convert a time interval into distance. I will briefly explain the logic behind this. Sound travels at 340m/s, which is 29 microseconds per centimeter. The ping needs to travel twice the distance (to the object and its rebound back to the sensor). Hence, we need to use 2*29 which is 58 microseconds per centimeter. This same logic is applied in the case of inches.

Now think about the maximum and minimum range. As seen in the above snippet, the maximum is set to 200 cm. Most hobby ultrasonic sensors can measure up to 200 cm without hassle, but this can be decreased according to your project. The minimum is set to 0cm because the sensor can indeed measure values at that distance but with lower accuracy.

In some cases, your Serial Monitor may be spamming 0cm as the sensor value, even though you know this is not the case. To fix this issue, simply replace `if (distance < 15)` with `if ((distance > 0) && (distance < 15))`.

Now that you have learnt how to use the ultrasound sensor, let's move on to the LCD part of the project.

Hooking up an LCD to the Arduino

The LCD screen that we will be using is an I2C LCD1602.

This display screen can be programmed to display whatever you want in a 16x2 matrix. This means that the screen (as you will soon find out) has two rows capable of fitting 16 characters in each row.

Before setting up the complete circuit, look at the back of the LCD. Plug in four wires, as follows:

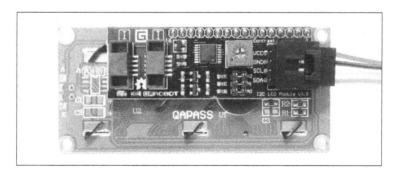

And then set up the circuit, as follows:

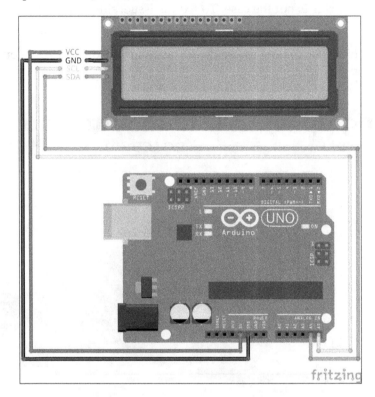

Now you will have to trust me on this next step. We are going to manually install a library that the LCD requires to run. You will be using this same method in future, so be patient and try to understand what we are doing here.

Download the `LiquidCrystal_I2C.zip` file from `http://www.wentztech.com/filevault/Electronics/Arduino/`.

Now, in the Arduino IDE, go to **Sketch | Include Library | Add ZIP library** and browse to the downloaded ZIP file. You are good to go.

If this doesn't work, you can manually extract the contents to: `C:\Users\<Username>\Documents\Arduino\libraries\LiquidCrystal_I2C` on Windows or `Documents/Arduino/libraries/LiquidCrystal_I2C` on the Mac and the same on Linux.

You will have to restart the IDE for it to be detected.

Now create a new sketch. Copy the following:

```
#include <Wire.h>
#include <LiquidCrystal_I2C.h>

LiquidCrystal_I2C lcd(0x20,16,2);  // set the LCD address to 0x20 for
a 16 chars and 2 line display

void setup()
{
  lcd.init();                      // initialize the lcd

  // Print a message to the LCD.
  lcd.backlight();
  lcd.print("Hello, world!");
}

void loop()
{
}
```

So few code lines after all that trouble? Yes! This is the beauty of using libraries, which enable us to hide all the complicated code so that the visible code is easier to read and understand. Save it as I2C_HelloWorld.ino. Plug in the Arduino and upload the code.

And you should have something like this:

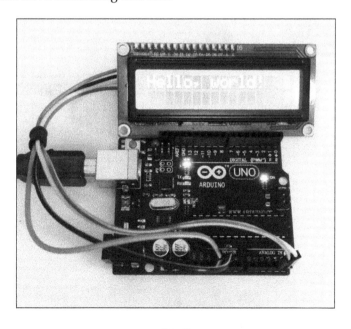

At this moment you are free to play with it, change the text from Hello, world! to whatever you like, such as "I like Arduino!".

But, what if I want to use the second line too? Well, don't worry. Change the code as follows:

```
#include <Wire.h>
#include <LiquidCrystal_I2C.h>

LiquidCrystal_I2C lcd(0x20,16,2);   // set the LCD address to 0x20 for
a 16 chars and 2 line display

void setup()
{
  lcd.init();                       // initialize the lcd

  // Print a message to the LCD.
  lcd.backlight();
  lcd.print("I like Arduinos!!!!!");
  lcd.setCursor(0,1);
  lcd.print("So Awesome!");
}

void loop()
{
}
```

Best of both worlds

Now comes the part where we combine what we have learnt so far in this chapter into one project. We are going use the sensor to calculate the distance between an object and relay this information to be displayed on the LCD screen in real time.

You are going to combine both the circuits from the previous two sections and create something like this:

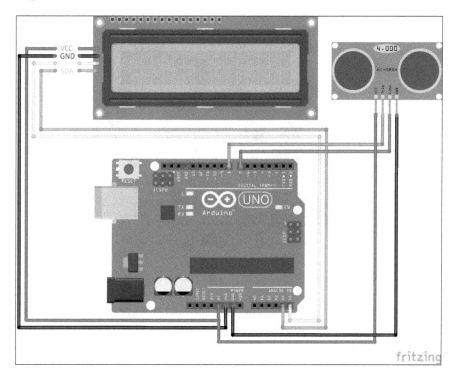

Note that, at the bottom of the setup, you can see two power VCC/5V/red wires merging into one. We did not use a breadboard because we wanted to save space. A simple way to go about this is to use a male splitter. A crude way is to cut a male to male in two and cut one female to female wire in two, strip off a bit of their plastic insulation and twist the copper ends (two males and one female) together and simply tape the joint.

Open up a new sketch and load the `Digital_Ruler.ino` file.

Save it as `Digital_Ruler.ino` in your `Chapter 2` directory and upload the code to the Arduino.

The result, if everything has gone right, will be exactly what you expect. It should look something like this:

Pretty neat, huh? But, if you want to make it compact, you can create something like this:

The LCD view will be like this:

Hold on a second! How did you get there? Well, as much as I would like to give you a step-by-step tutorial for putting the setup into a box, I'd like to take this moment to ask you to put on your creativity cap, scramble around your home/office, and find scraps that you could use to convert this into an art and craft project. You can use anything you can find: a box, tin can (insulated of course), a bottle, anything would do. I merely chose the packaging box that the LCD came in.

But what if I want to make it completely portable? As in, without that annoying USB cable? Well, let me show you what you can do. You simply need to have these two things:

- A 9V Battery:

- A 9V battery connector:

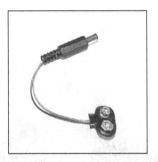

And build the following circuit:

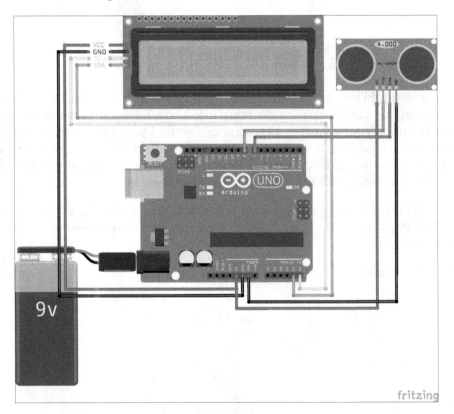

The code is the same. Everything remains the same except that there is a battery to power the board instead of the USB cable. A standard 9V battery connected to a 2.1mm barrel jack can be connected to the Arduino to power it. The Arduino UNO can handle between 5-20V of voltage. But the onboard voltage regulator ensures that no more than 5V is fed to the components connected to it. If you want to, you can also mount a laser pointer on the sensor so that it improves the accuracy of the device. Plus, it would look a lot cooler.

Summary

That was fun, right? And a bit challenging, correct? Good! That's when you know you are learning. So let's just summarize what we achieved in this chapter. We first programmed a HC-SR04 Ultrasound sensor and used it to measure a distance which was then displayed on the Arduino UNO Serial Monitor. Next, we played around with the I2C LCD1602 screen, and then, we combined what we learned from the two sections into one project called the Digital Ruler (Scale). You successfully created a digital measuring tape, which made it compact and less cumbersome to use. But since it can measure between 0 to 2 meters, it can only be used indoors. Higher ranges can be achieved using better (and more expensive) sensors.

In the next chapter, we will learn about touch sensors, which will along with a powerful processing software allow us to convert finger gestures to text.

3
Converting Finger Gestures to Text

You have reached chapter 3. Here we will look deeper into the realm of sensors. You will learn a lot in this chapter about using a touch sensor to create cool projects. You will be introduced to Arduino's sister software (technically its father). Processing, which is often used along with the Arduino to either create an interface to communicate with the Arduino, or simply to display data in a much more comprehensible way.

This chapter is divided into three sections. We will start by learning the basics of Processing. Then we will use the number touch pad and the processing software to create a tic-tac-toe (X's & O's) game. And, in the final section, we will use the touch pad to recognize stroke patterns and different alphabets and display them on the screen.

This chapter uses less hardware; however it involves a lot of programming, so you should be ready for that.

In this chapter, we'll cover the following topics:

- Brief note on capacitive sensors
- Introduction to Processing
- Tic-tac-toe with touch
- Pattern recognition with touch

Prerequisites

You will need the following components to add a touch element to your Arduino projects:

- 1x Arduino-compatible board such as the UNO
- 1x USB Cable A to B
- 1x capacitive touch kit (`http://www.dfrobot.com/index.php?route=product/product&keyword=capac&product_id=463`)

[Only the capacitive number pad and touch pad will be used in this chapter.]

What is a capacitive touch sensor?

A capacitive touch sensor is an upgrade from the commonly used resistive sensors that relied on a change in resistance due to a change in resistor geometry due to the applied pressure. A capacitive sensor works on the principle that your finger acts as an external capacitor that changes the total capacitance of the system, which is then measured by onboard chips and converted into readable data.

An introduction to Processing

Processing is very similar to Arduino in the sense that it has a similar interface to the Arduino IDE. But Processing is used mainly to create graphical interfaces or to display data in real time, which is what we are going to be doing in this chapter. Let's get started.

The first thing you will have to do is to download and install the latest version of Processing from `https://processing.org/download/`.

Install it like you installed Arduino.

Open the application. The environment looks very similar to the Arduino IDE, right? This should make it easy for you to work with.

For a quick demo on what Processing can do, go to **File | Example** and, under **Inputs**, choose **MouseXY (Mouse2D)**. Run the sketch and be amazed! Try some other examples, such as (**Topics | Simulate | Flocking**), to see what else can be achieved.

Now we are going to use this powerful software along with the Arduino and number touch pad to create a tic-tac-toe game.

Tic-tac-toe with touch

Remember how, while using Arduino, you needed to install libraries to make certain functions work? We need to do the same with Processing because it cannot directly communicate with the Arduino. To do this, go to `http://playground.arduino.cc/Interfacing/Processing` and download the `processing2-arduino.zip` file. Processing, like Arduino also creates a directory by default in the `Documents` folder. Extract the downloaded ZIP file to `C:\Users\<user>\Documents\Processing\libraries` for Windows and `Documents/Processing/libraries` for Mac and Linux.

Do the same for the matrix library we will be using in the next section from `http://pratt.edu/~fbitonti/pmatrix/matrix.zip`. **Restart** Processing.

 If you do not have a `libraries` folder in the `Processing` directory, go ahead and create a new one.

Now launch the Processing IDE (not the Arduino IDE). Connect the Arduino and run this sketch:

```
import processing.serial.*;
import cc.arduino.*;

Arduino arduino;
int ledPin = 13;

void setup()
{
  //println(Arduino.list());
  arduino = new Arduino(this, Arduino.list()[0], 57600);
  arduino.pinMode(ledPin, Arduino.OUTPUT);
}
```

```
void draw()
{
  arduino.digitalWrite(ledPin, Arduino.HIGH);
  delay(1000);
  arduino.digitalWrite(ledPin, Arduino.LOW);
  delay(1000);
}
```

What do you observe? The LED that is connected to pin 13 should begin blinking. This means that you have coded the Arduino successfully with Processing. If it didn't work, try using `Arduino.list()[1]` instead of `Arduino.list()[0]`.

Ok, now we want to test out the capacitive touch pad. Create the following circuit, taken from DFRobot's wiki:

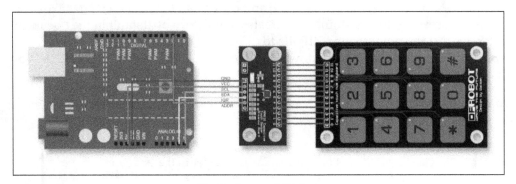

(Credits: Lauren, DFRobot)

The connections are as follows (Arduino UNO | NumPad):

- GND – GND
- VCC – 5V
- SCL – A5 (analog pin 5)
- SDA – A4
- IQR – D2 (digital pin 2)
- ADDR – no connection necessary

This particular touch pad also requires a library. Go to `http://www.dfrobot.com/image/data/DFR0129/MPR121%20v1.0.zip` to download it. As before, extract it to the `Processing libraries` directory.

Now it is time to get to the actual tic-tac-toe program.

Arduino and Processing

Open up a new sketch on Arduino and paste in the following:

```
#include <Wire.h> // default Wire library
#include <mpr121.h> // touch pad library

int num = 0; // variable to store pressed number

void setup()
{
  Serial.begin(19200); // begin the Serial Port at 19200 baud
  Wire.begin(); // initiate the wire library
  CapaTouch.begin(); // inititate the capacitive touch library
  delay(500);
}

void loop()
{
  num = CapaTouch.keyPad(); // stores the pressed number to num
  if(num > 0){ // checks if the key pressed is within scope
    Serial.print(num); // prints the number to the serial port
  }
  delay(200); // small delay to allow serial communication
}
```

Save it as before, to `tic_tac_toe.ide`.

You must be thinking: Oh! You said there would be a lot of programming in this chapter! This code is so small! Running the program will give you a **Serial Monitor** display like this:

Serial Monitor output

Well, that was just the Arduino sketch. Now we need to add the processing sketch that will allow us to communicate with the Arduino and create a nice little graphical interface where you can play a two-player tic-tac-toe game.

This involves a lot of things, as you will find out.

Using Processing, open the file called `tic_tac_toe_pro.pde` that came with this chapter and, when you're ready, run it.

The result

Go ahead and play a game on your own or with a friend and you should get something like this:

Tic-tac-toe final output

Pretty neat, huh? If you are feeling really adventurous and have complete confidence in your programming skills, you can go ahead and program a player versus AI game, which is outside the scope of this book.

Now that we have had a small taste of what is possible using a touch sensor, we will move on to the pattern recognition part of the chapter where we will push the capacitive grid sensor and our coding capabilities to their limits.

Pattern recognition

Firstly, go ahead and create this circuit:

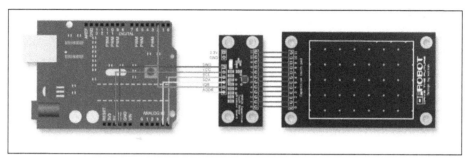

Touch pad circuit (credits: Lauren, DFRobot)

Your capacitive touch pad and its controller (the central component in the image), are connected using the corresponding numbers labeled on the pins. The connections from the Arduino to the controller are as follows (literally the same connections as before):

- GND – GND
- VCC – 5V
- SCL – A5 (analog pin 5)
- SDA – A4
- IQR – D2 (digital pin 2)
- ADDR – no connection necessary

Do not worry if the touch pad doesn't look exactly like this; as long as the connections are fine, everything will work out.

Open up Arduino and go to **File | Examples | MPR121 | Examples | Touchpad** or copy the following code:

```
#include <Wire.h>
#include <mpr121.h>

int X ;           // X-coordinate
int Y ;           // Y-coordinate

// =========  setup  =========
void setup()
{
  //  initialize function
```

```
    Serial.begin(19200);
    Wire.begin();
    CapaTouch.begin();

    delay(500);
    Serial.println("START");
}


// ========= loop =========
void loop()
{
  X=CapaTouch.getX();                    // Get X position.
  Y=CapaTouch.getY();                    // Get Y position.
  if(X>=1 && X<=9 && Y>=1 && Y<=13)
    {
// Determine whether in the range. If not, do nothing.
      Serial.print("X=");
      Serial.print(X);
      Serial.print("  Y=");
      Serial.println(Y);
    }
  delay(300);
}
```

Now, run this program, open up the Serial Monitor, and set it to a baud rate of 19200. Play around with the touch pad using your finger tip and you should get something like this:

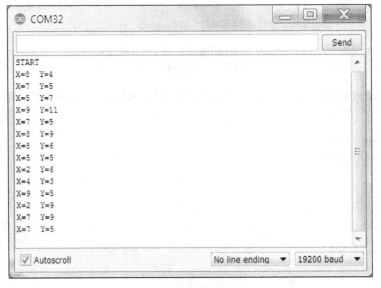

Serial Monitor with touch pad

Before we jump into how we are going to use this device, let's take a minute to think about what exactly we are trying to do using the device. You can use it for various things but, for the purpose of this project, we are going to use the touch pad, Arduino, and Processing to convert your finger strokes into meaningful text (alphabets for simplicity).

Now, before we link this to Processing, we need to learn how to tackle the problem that has haunted engineers for decades, namely, image processing. In this case, however, the image processing is not too complex, but some sort of logical brainstorming is required. Look at your capacitive touch pad.

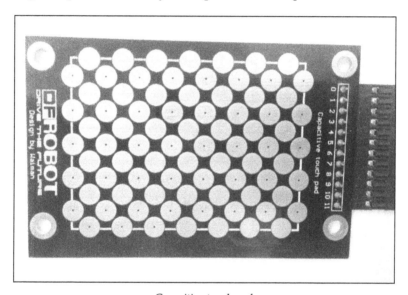

Capacitive touch pad

What is the first thing that you think of that would aid in analyzing the stroke patterns? Right away, you would guess, a two-dimensional array! You are right.

A 2D array or matrix like the one that we used in the X's & O's section will help us to tackle this problem. If you have already played with the touch pad, and if yours looks similar to the one used in this chapter, the first two rows both have an X value of 1, the last two have a value of 13, the two left columns have a Y value of 1 and the two right-most columns have a Y value of 9.

Let's try to put them to good use. This is what the layout of the touch pad would look like with coordinates added in for visualization.

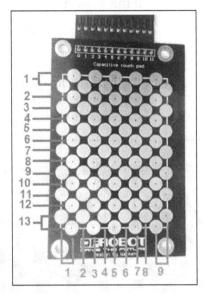

Capacitive touch pad - Labeled

We are going to create a 13x9 matrix with a starting value of 0 for each element. When the capacitive touch sensor pad is touched, the corresponding value in the matrix will be changed to 1. In this way, we have a mathematical means of representing the stroke layout.

To further alleviate the complexity of the problem of having to understand and iterate all of the twists and bends of alphabets, we will be using block diagrams.

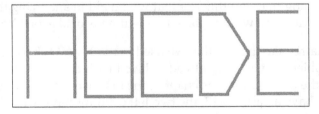

Block diagram for ABCDE

What is wrong with that D? It is enough to know that, if the D was written as a complete rectangle, it would simply represent O. These typefaces are much easier to process as they are mainly composed of lines.

Let's look at the block letter A.

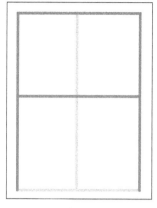

Block A

Let's name each of the straight lines:

- Top Horizontal – T
- Bottom Horizontal – B
- Middle Horizontal – H
- Left Vertical – L
- Right Vertical - R
- Middle Vertical - V

With the labels, it looks like this:

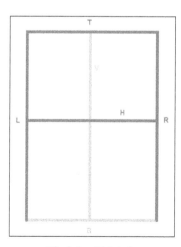

Block A with labels

Note how the Bottom Horizontal line and the Vertical Line are faded; this is because the alphabet A does not contain these strokes. We can calculate if a particular combination of T, B, H, L, R, and V exists for each alphabet in the matrix we made before. If we denote true (exists) as 1, then for A:

- T = 1
- B = 0 (does not exist)
- H = 1
- L = 1
- R = 1
- V = 0

You see where we are going with this, right? In this way we can map out the characteristics of all the alphabets. Some alphabets, though, are a bit tricky. For example, the letter P looks like this:

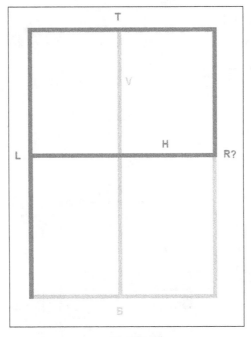

Block P

What do we do for the Right Vertical (R) line? To solve this issue, if we come across a half-line, we will denote it with a 2. A ½ would be more representative but, when using ½ as a coding parameter, it will simply disappear into a zero in the integer format, hence 2 is used.

So P would become:

- T = 1
- B = 0
- H = 1
- L = 1
- R = 2
- V = 0

Now, to represent the touch panel in real time or to process a display to represent it requires the ability to use the power of matrices. So we will code a simple layout that looks like this:

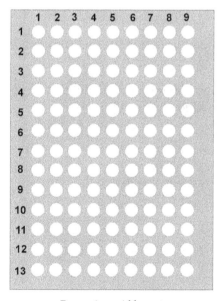

Processing grid layout

And, with the strokes on an ideal block A, it would look something like this:

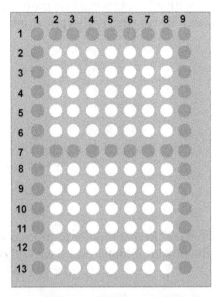

Processing grid layout for 'A'

Now that we have understood how to solve this entire problem, let's finally go to the code.

Fire up Arduino and paste in the following code. Or you could just use the touch_pad.ide file supplied to you with this book.

```
// Connections:
// SDA (MPR121) -> PIN A4 (Arduino)
// SCL (MPR121) -> PIN A5 (Arduino)
// IRQ (MPR121) -> PIN A2 (Arduino)

// Libraries
#include <Wire.h>
#include <mpr121.h> // Touch Pad library

int X ;   // X-coordinate
int Y ;   // Y-coordinate
int inByte = 0; // incoming serial byte

void setup()
{
```

```
  Serial.begin(57600); // Begin serial at 57600 baud rate (faster)
  Wire.begin(); // intiialize wire library
  CapaTouch.begin(); // initialize the capacitive touch library
  delay(500); // brief delay for initialization
}

void loop()
{
  X = CapaTouch.getX(); // Get X position.
  Y = CapaTouch.getY(); // Get Y position.

  // Determine whether in the range.If not,do nothing.
  if(X>=1&&X<=9&&Y>=1&&Y<=13)
  {
    // Debug lines, can be uncommented to check inputs
    //Serial.print("X=");
    //Serial.write(X);
    //Serial.print("   Y=");
    //Serial.write(Y);

    // convert X and Y coordinates into one variable
    if(Y<10)
    {
      Serial.print(X*10+Y); // prints to serial port
    }
    else if(Y>9)
    {
      Serial.print(X*100+Y); // prints to serial port
    }
  }
    delay(200); // delay for message to be relayed
}
```

Save it as touch_pad under Chapter 3.

Note how we converted both the X and Y coordinates into one integer so that we send a single data point through the serial port.

Open a new processing sketch and open the file named touch_pad_pro.pde. Run it.

Save it as touch_pad_pro. Upload the Arduino code first and only then run this program! If you get an error with the libraries, make sure they are installed in your Processing directory in Documents.

Go ahead and play with it. I recommend the first character to be an 'I', nice and simple. Make sure the strokes are gentle. You can go over previous lines, so do not worry about that. You will soon find out that the strokes don't have to be perfect for the program to identify what they are.

Your results should look identical or similar to these:

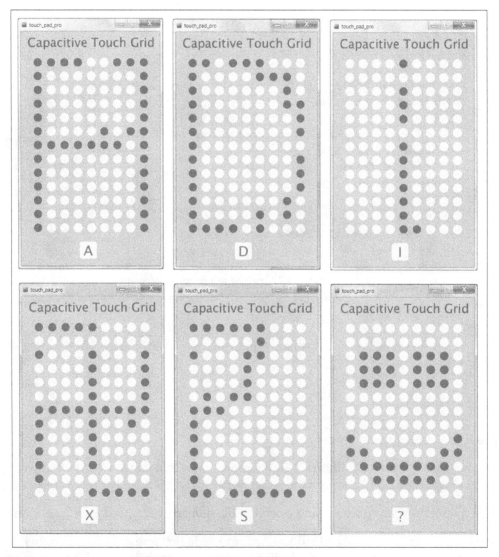

Touch pad results

With this knowledge you can do almost anything with that touch device. Think maze game, digital locking mechanism, and the like.

Summary

That was super cool, wasn't it? The synergy when Arduino, Processing, and touch sensors work together is enormous. We first played with the basics of Processing and understood how it communicates with Arduino. We wrote a simple Processing code that could directly control the Arduino. Then we created a simple tic-tac-toe program using the capacitive number pad as the input device, and using Processing as the processor to make the game tick. Finally, we created a complicated pattern-processing program that takes input from the capacitive touch pad, which we realized is really powerful. You have learnt a lot from this chapter and I hope this serves you well in your coding endeavors.

In the next chapter, we bid farewell to Processing and say hello to an even more powerful tool that synergizes well with Arduino, namely, Python. Additionally, in the next chapter, we will learn how to implement wireless communication in our projects by means of Bluetooth.

4
Burglar Alarm – Part 1

In the previous chapter, there were a lot of coding skills to take on board. Understandably, it was difficult, but you made it through. In this chapter, we will use a lot of hardware and software. Haven't you always wanted to make your own burglar alarm? Catch the crook who stole your favorite cookies from the cookie jar? This chapter will teach you how to do just that.

Firstly, we will create a plan of how we are going to go about catching the culprit. Of course, it is not enough to simply sound an alarm when a thief is caught in the act; you also want to have evidence. You see where this is going, right? Yes, we will be using a wireless camera to get a mugshot of the culprit.

This chapter does use a fair share of code but, in addition to that, it requires quite a few components. Learning to use them in unison is the ultimate goal of this chapter. It is divided into the following sections:

- The PIR sensor
- Testing the camera
- Communicating with a smart phone
- The burglar alarm

I promised you in previous chapters that I would try to teach you as much I possibly could about the Arduino. Bluetooth is very reliable and inexpensive but it is short-range and usually needs a host to gather or send data.

The following are the components you'll need, to create a high-tech burglar alarm:

- 1 x Arduino UNO board
- 1 x USB cable A to B (aka the printer cable)
- 1 x PIR sensor
- 1 x wireless IP camera (a netcam360 is used in this chapter)
- 1 x HC-06 module (Bluetooth)
- 1 x wireless router (with accessible settings)
- 1 x PC with inbuilt Bluetooth or a Bluetooth USB module

What is a passive infrared sensor?

A **passive infrared sensor** (**PIR**) is an electronic sensor that uses infrared radiation to detect variations in its field of view. They are most commonly used as motion sensors; for example, they are used to minimize power consumption by switching off lights and utilities if nobody is at home. They are also used in state-of-the-art burglar alarm systems to trigger a switch when motion is detected.

 If you would like to learn more about how they work, you should refer to this page (https://learn.adafruit.com/pir-passive-infrared-proximity-motion-sensor/) at Adafruit. Adafruit and Ladyada are really good resources for building Arduino projects.

A mini PIR-Arduino alarm

Let's get started. We are going to create a setup in which an LED flashes when motion is detected by the PIR sensor. This is what the setup should look like when connecting the Arduino to the PIR Sensor:

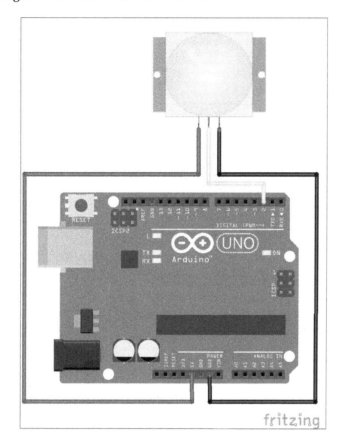

Basically, the connections are as follows:

- GND → GND
- VCC → 5V
- OUT → D02 (digital pin 2)

Digital pins are denoted using D and analog pins are denoted by A.
So digital pin 13 is D13 and analog pin 2 is A02.

Open Arduino and load the sketch called `PIR_LED.ino`, or copy this:

```
int ledPin = 13; // use the onboard LED
int pirPin = 2; // 'out' of PIR connected to digital pin 2
int pirState = LOW; // start the state of the PIR to be low (no
motion)
int pirValue = 0; // variable to store change in PIR value

void setup() {
  pinMode(ledPin, OUTPUT); // declare the LED as output
  pinMode(pirPin, INPUT);  // declare the PIR as input
  Serial.begin(9600); // begin the Serial port at baud 9600
}

void loop() {
  pirValue = digitalRead(pirPin); // read PIR value
  if ((pirValue == HIGH)&&(pirState==LOW)) { // check if motion has
occured
    digitalWrite(ledPin, HIGH); // turn on LED
    Serial.println("Motion detected!");
    pirState = HIGH; // set the PIR state to ON/HIGH
    delay(1000); // wait for a second
  }
  else { // if there is no motion
    digitalWrite(ledPin, LOW); // turn off LED
    if(pirState == HIGH) {
    // prints only if motion has happened in the first place
    Serial.println("No more motion!\n");
    pirState = LOW; // sets the PIR state to OFF/LOW
    delay(200); // small delay before proceeding
    }
  }
}
```

Run the code and open up the **Serial Monitor** and set the baud rate to 9600.
This is a simple program that switches on the LED when motion is detected and
powers it off when there is no motion.

See it in action by moving your hand in front of the PIR censor. See how the LED glows when the PIR detects motion? PIR sensors are very sensitive to variations in light, which is why they are often used in motion detectors.

This was pretty straightforward so far, right? Good! Let's now move on to the camera.

Testing the camera

An IP camera (or Internet Protocol camera) is a camera that you can access on your wireless network provided that it is configured correctly. The configuration procedure depends on which IP camera you bought and who the manufacturer is, but they should all be pretty similar.

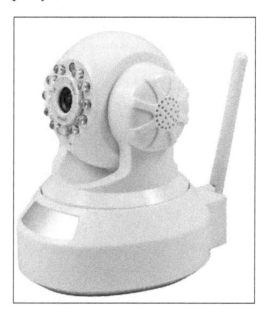

If you purchased the exact same one that is used in this chapter, the setup is explained in the following section. However, if you got a different one, do not worry. Use the manual or installation guide that comes with the camera to install it onto your network. Go through the following steps anyway, to get an idea of what you should do if you do not have the same camera.

Installing the camera on the network

Go to `http://netcam360.com/enindex.html` and download **IP camera for PC**.

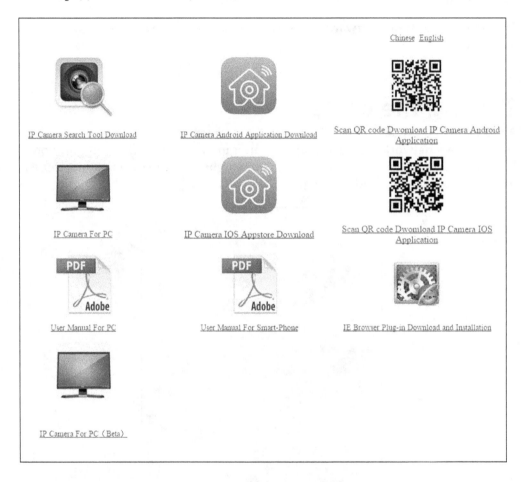

You should also be able to find a PC installation manual on the same webpage. Follow the instructions provided. I will not go into detail because there is a high chance that you have not purchased the same IP Camera.

Eventually, you should be able to get something like this:

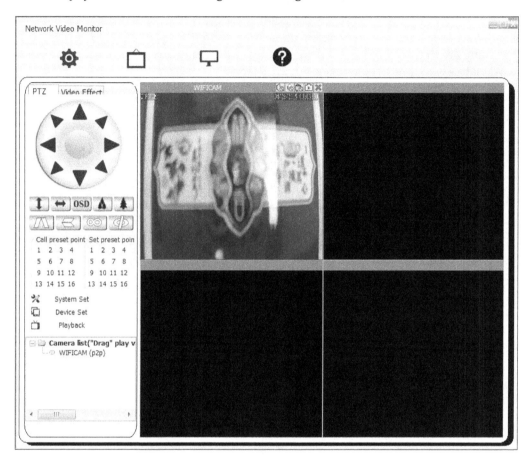

It is highly recommended that you create a password for your camera to keep prying eyes away. Also, do remember the password because we will need it in the steps to come.

Setting up the mugshot URL

Go back to `http://netcam360.com/enindex.html` and download the IP camera search tool or use the file of the same name located in the `Useful Files` folder that comes with this chapter. I believe that this tool can be used, irrespective of the manufacturer, to find IP cameras on your network.

Run it and you should see this screen:

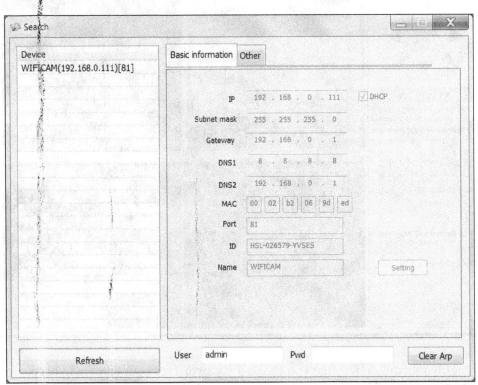

Note down the IP address at the top. In this case, it is 192.168.0.111:81, where 81 is the port number.

Open a new browser window and, in the address bar, paste the following:

`http://<your ip>/snapshot.cgi?user=admin&pwd=<your_password>`

For example:

```
http://192.168.0.111:81/snapshot.cgi?user=admin&pwd=password
```

You should be able to see a snapshot of what your camera is seeing in real time. You can try refreshing the page while moving the camera to test different camera positions. We are going to use this link later in the chapter to fetch the image file to catch the burglar.

Putting it together

As I mentioned before, we will be using a Bluetooth module to communicate with the Arduino. The specific model is the HC-06 module, which is one of the cheapest Arduino communications modules and is widely used. We could just try to hook it up to our smart phone and send an alert directly to it, but what if you aren't at home? That is a drawback of using Bluetooth.

But fret not. We will capitalize on what we can do using just the Bluetooth. I think this is a good time to explain the plan before diving into the details.

The following is a representation of what we are going to achieve in this part of the chapter.

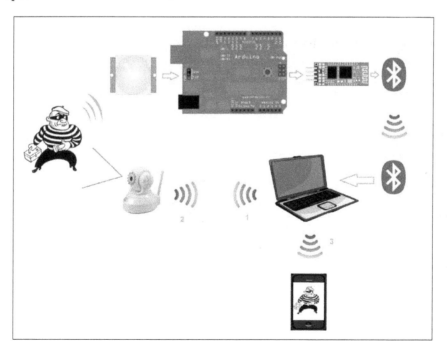

We use the PIR sensor to check if there is any motion detected. This data is transmitted through the HC-06 Bluetooth module to a host computer through a Bluetooth channel. A script (program) running on the computer takes a snapshot of the culprit using the wireless IP camera's URL and saves it as an image. It then uploads the image to an image-sharing website and fetches the link. Finally, an alert is sent to the user saying that the alarm has been set off. The user can then open the link on a smart phone to see the culprit. I hope that wasn't too hard to follow.

First, let's get the script ready.

An introduction to Python

Wait! What? I did not sign up for this! Hold on there. We are not going to study Python too much. That is outside the scope of this book. To be honest, I think this particular section is the furthest you will deviate from the field of Arduino. But you must understand that, to create something really powerful, you need to make the most out of the resources we have. We are just going to use its basic functionalities to achieve our intended plan.

What are you even talking about? What is Python? Why are we talking about a snake? Python is a very powerful, but easy to use, language like C or Java. We will be using it to get the snapshot from the camera, upload it to the Web, and send a notification to your smart phone.

The following image is the best way to describe it:

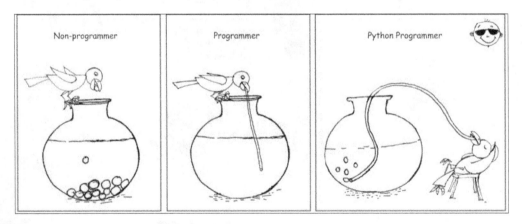

You should be warned that this section is going to be a bit difficult, but you should try to be patient. Once Python is installed, it is going to stay there forever. Let's go ahead and install it. There are two popular versions of Python, namely 2.7 and 3.4. We are going to use 2.7 since it is older and has a lot more libraries that work with it. Yes, libraries, similar to the ones that you used earlier with the Arduino.

Download Python from `https://www.python.org/download/releases/2.7.8/` according to your operating system. Note that 32-bit is x86 and 64-bit is x64.

Install it to somewhere convenient. Most OS X and Linux computers come with Python pre-installed.

Next we need to download and install an interface for writing your codes. I recommend Eclipse because it is easy to use for newcomers to Python. Since Eclipse is Java-based, you should update Java by going to `http://www.java.com/en/` and installing/updating Java on your system.

You can download Eclipse from `http://www.eclipse.org/downloads/`. Select Eclipse Classic and install it, or rather extract it, into a convenient location, as with Arudino. If you have used Python before, you can simply choose your own interface. Geany is one commonly used Python IDE.

Open Eclipse. If you get a Java error, use this resource to solve the problem: `http://stackoverflow.com/questions/2030434/eclipse-no-java-jre-jdk-no-virtual-machine`. It will ask you for a workspace. Use something like `C:\MyScripts` or choose anything similar. Then close the welcome screen. You will see something like this:

Ignore **PyDev Package Explorer** on the left in the screenshot. I am starting from the beginning so that you can set it up.

1. Go to **Help | Install New Software**.

   You will see this dialog box:

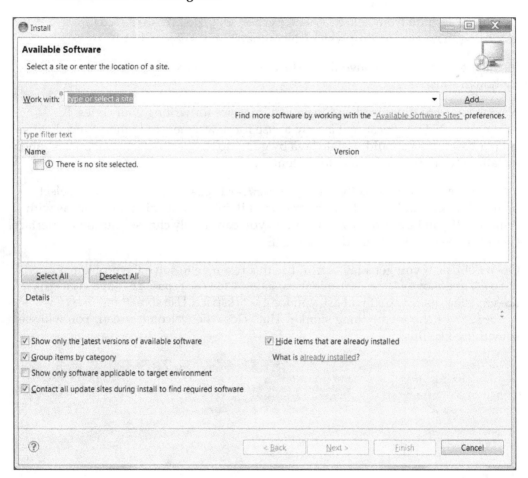

2. Click **Add...**.

3. For **Name**, type pydev and for **Location,** type http://pydev.org/updates.
4. Press **OK**. Wait for it to load.

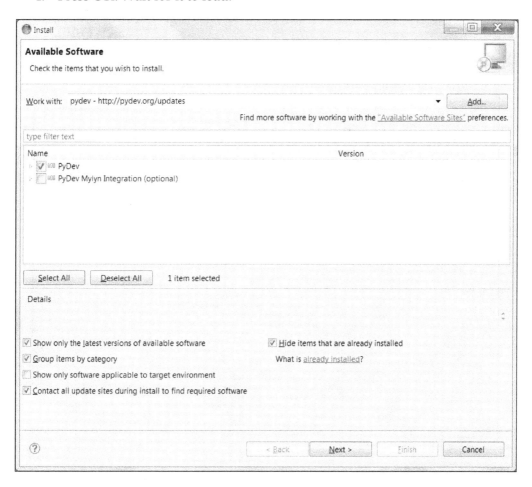

5. Select the first option (**PyDev**) and click **Next >**. Accept the terms and conditions and let it install. If it asks you whether you trust this application, just select **Yes**.

   It will ask you to restart Eclipse. Allow it. Wait for it to launch by itself. You're almost done.

6. Go to **Window | Preferences**.

   In that dialog, expand **PyDev | Interpreters | Python Interpreter**.

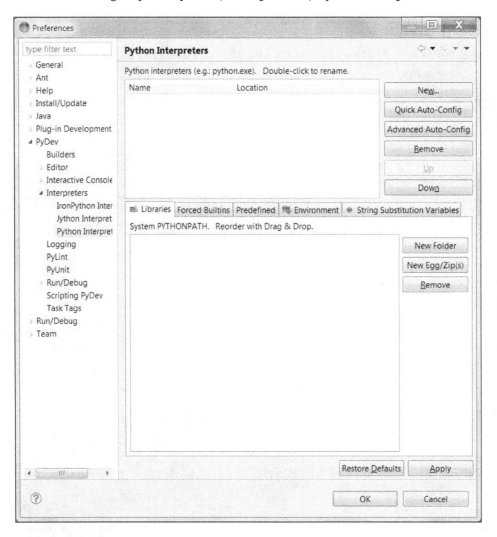

7. Click on **New…**.

8. For the **Interpreter Name**, type `Python27`.

    And for **Interpreter Executable**, you can browse to select `C:\Python27\python.exe` or you can simply paste that without quotations. Click **OK**.

    It will load a lot of files. Ensure all are checked and hit **OK**.

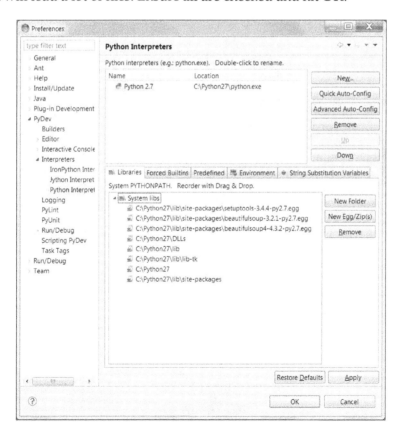

9. Hit **OK** again.

   One last thing: on the Eclipse Interface on the top right. Select **PyDev**
   instead of Java.

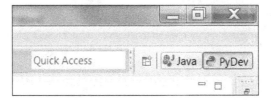

10. Now, in **PyDev Package Explorer**, right-click and create a new project.

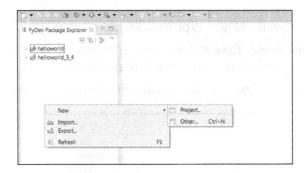

11. Choose **General | Project**.

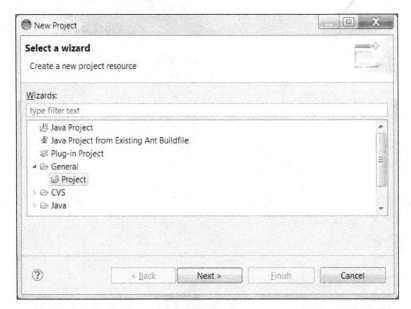

12. Click **Next >**.

Call your project something like **MyProject** or **helloworld**, or whatever you like, and let it use the default location to store the files. Click **Finish** when you're ready.

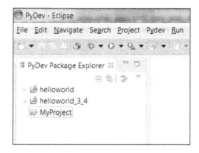

13. Right-click on **MyProject | New | File**.

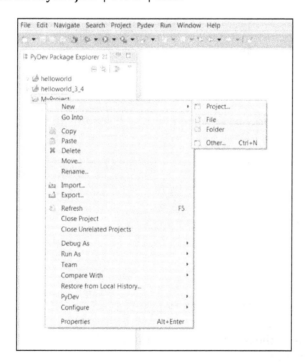

14. Call it `helloworld.py` and press **Finish**.

Remember to use `.py` and not just `helloworld`, so that Eclipse understands what to associate it with. Type:

```
print("Hello World!")
```

15. Press **Run** (the green bubble with a right-pointing triangle). It will bring up the following window every time you run a new script:

16. Simply select **Python Run** and press **OK**.

    If everything works as expected, you will see this in the **Console** window on the right side:

Hurray! You have just written your very first Python script! You should be proud of yourself, because you have just been introduced to one of the most powerful programming tools in the history of programming. When you are done with this chapter, I recommend you look up some Python tutorials to see what it can do.

But now, we must keep moving. Let's now focus on setting up the Bluetooth network.

Hooking up the Bluetooth module

The HC-06 is a simple, low-powered Bluetooth module that works very well with Arduinos.

Go ahead and create this circuit:

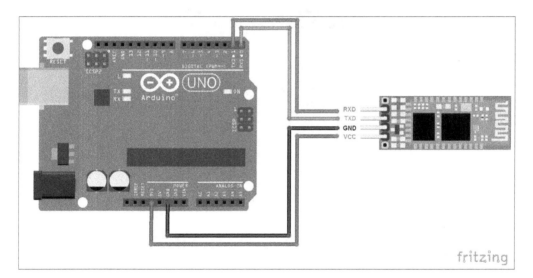

The connections are as follows:

- GND → GND
- VCC → 3.3 V
- RXD → D01 (TX)
- TXD → D00 (RX)

Note that, before uploading codes where RX and TX pins are used, unplug those pins. Reconnect them once the uploading process is complete.

1. Plug the Arduino into the USB hub to power the HC-06 chip. Now, in your system tray, right-click on the Bluetooth icon and click **Add a Device**.

2. Let the computer search until it finds **HC-06**.

If nothing shows up, try using an Android phone to connect to it. If it doesn't show up even then, check your connections.

3. Click **Next**.

   Here, type in the device's pairing code, which is 1234 by default.

4. It will now install the HC-06 on your computer. If everything works well, when you open up **Device Manager** and go to **Ports (COM & LPT)**, you should see this screen:

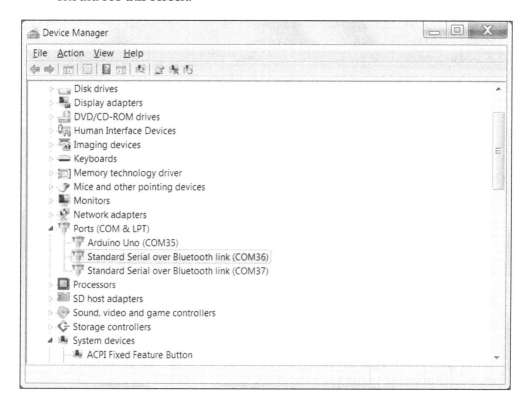

   Note down these three COM values (they will be different for different users).

5. Finally, you are ready to program the Bluetooth module.

Open up a new Arduino sketch and load the `ard_BL_led.ino` file or paste in the following code:

```
char bluetoothVal;          //value sent over via bluetooth
char lastValue;             //stores last state of device (on/off)

int ledPin = 13;
```

```
void setup() {
 Serial.begin(9600); // begin communication on baud 9600
 pinMode(ledPin, OUTPUT); // set the led pin to output
}


void loop() {
  if(Serial.available()) // searches for available data
  {//if there is data being recieved
    bluetoothVal=Serial.read(); //read it
  }
  if (bluetoothVal=='1')
  {//if value from bluetooth serial is '1'
    digitalWrite(ledPin,HIGH);      // turn on LED
    if (lastValue!='1')
      Serial.println(F("LED ON")); //print LED is on
    lastValue = bluetoothVal;
  }
  else if (bluetoothVal=='0')
  {//if value from bluetooth serial is '0'
    digitalWrite(ledPin,LOW);       //turn off LED
    if (lastValue!='0')
      Serial.println(F("LED OFF")); //print LED is off
    lastValue = bluetoothVal;
  }
  delay(1000);
}
```

Again, before uploading, make sure you disconnect the RX and TX wires. Connect them after the upload is completed.

To test this code, we will use some popular software called Tera Term. OS X and Linux systems come with terminal emulators so this is not necessary. It is mainly used as a terminal emulator (a fake terminal, in plain language) to communicate with different devices/servers/ports. You can download it from http://en.osdn.jp/projects/ttssh2/releases/. Install it to someplace convenient.

Launch it and select **Serial**, and select the COM port that is associated with Bluetooth. Start by using the lower COM port number. If that doesn't work, the other one should.

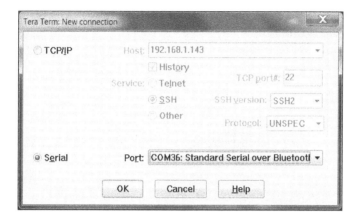

Hit **OK**. Give it some time to connect. The title of the terminal window will change to **COM36:9600baud** if everything works correctly.

Type 1 and hit enter. What do you see? Now try 0.

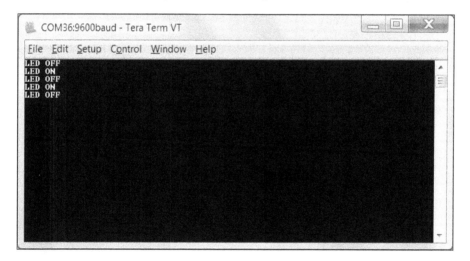

 Give Tera Term some time to connect to Bluetooth. 1 or 0 are not displayed when you type them. Just the LED status will be displayed.

You have now successfully controlled an LED via Bluetooth! This effect would be a lot cooler if you used a battery to power the Arduino so that there was no wired connection between the Arduino and the computer. Anyway, let's not get carried away, there is much to be done.

Before bringing everything together, there are two things left to be done: dealing with the image (mugshot) upload and sending a notification to your smart device. We'll start with the former, in the following chapter.

Summary

No, we are not done with this project. The second half of it is moved to the next chapter. But let's do a quick recap of what we have done so far. We tested out the PIR sensor, which we found out to be a really efficient motion sensor. We installed and wrote our very first Python script, which is a phenomenal achievement. Finally, we used Bluetooth to communicate between the computer and the Arduino.

In the next part of this project, we are going to process the image captured from the camera, upload it to where it can be accessible on other devices, learn about notification software, and finally bring the pieces together to create the burglar alarm.

Burglar Alarm – Part 2

5

This is part 2 (and the final part) of the burglar alarm series. So far, we have configured the camera, the Bluetooth, and Python.

In this chapter, we will be going through the following topics:

- Obtaining and processing the image of the intruder
- Uploading the image to a convenient website
- Sending the URL to your smart phone

So, shall we get right to it?

Dealing with the image

As discussed before, when Arduino sends a message to Python, it is going to take a snapshot using the camera and save it to the computer. What do we do with it, then? How do we upload it to a file sharing platform? There are several ways to do this, but in this chapter, we will be using **Imgur**. Yes, the same Imgur that you have been using, knowingly or unknowingly, on Reddit or 9gag.

Go to `http://imgur.com/` and sign up for a new account.

Once you have verified your account, go to `https://api.imgur.com/oauth2/addclient` to add a new application so that Imgur permits you to post images using a script. Use the following information:

- **Application name**: Arduino burglar alarm
- **Authorization type**: OAuth 2 authorization without callback URL
- **Email**: <your email>
- **Description**: Using Arduino to catch the cookie thieves

Now proceed, and you will get a page like this:

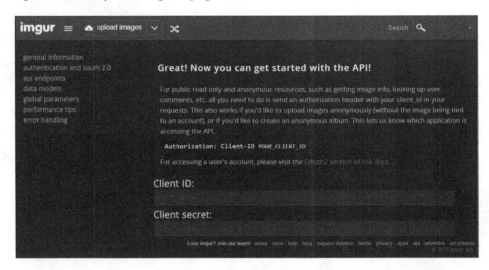

Imgur will give you the **Client ID** and **Client secret**. Save them in a safe location as we will need them later.

Let us try testing the automatic image uploading functionality. However, before that, we need to install the Imgur library for Python. To do this, firstly, we need to install a simple Python library installer called `pip`, which is used to download other libraries. Yeah! Welcome to Python!

Go to `https://pip.pypa.io/en/latest/installing.html` and download `get-pip.py`, or you could just use `get-pip.py` that is in the `Useful Files` folder that came with this chapter. Save it or copy it to `C:\Python27`. Go to your `C` directory and *Shift* + right-click on `Python27`. Then, select **Open command window here**:

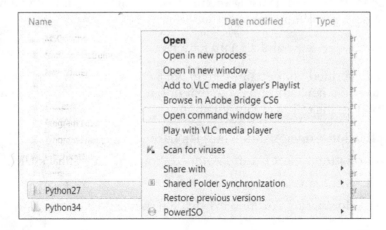

Type `python get-pip.py`, hit *Enter*, and let it install pip to your computer.

Navigate to `C:\Python27` and *Shift+* right-click on **Scripts**. Open the command window and type `pip install pyimgur` in that command window to install the Imgur library. Your terminal window will look like this:

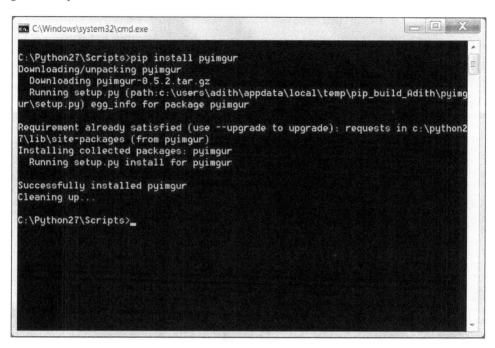

Let's test this newly installed library. Open Eclipse and create a new file called `imgur_test.py`, and paste the following or simply load the `imgur_test.py` file from the code attached to this chapter:

```
import pyimgur # imgur library
import urllib  # url fetching default library
import time    # time library to add delay like in Arduino

CLIENT_ID = "<your client ID>" # imgur client ID

# retrieve the snapshot from the camera and save it as an image in the
chosen directory
urllib.urlretrieve("http://<your ip : port>/snapshot.
cgi?user=admin&pwd=password", "mug_shot.jpg")
time.sleep(2)
```

```
PATH = "C:\\<your python project name>\\mug_shot.jpg" # location where
the image is saved

im = pyimgur.Imgur(CLIENT_ID) # authenticates into the imgur platform
using the client ID
uploaded_image = im.upload_image(PATH, title="Uploaded with PyImgur")
# uploads the image privately
print(uploaded_image.link) # fetches the link of the image URL
```

Change `<your client ID>` to the ID that you saved when you created an application on Imgur; change `<your ip : your port>` to your camera URL; change `<your python project name>` to the folder where all your python codes are saved. For example, if you look at your Eclipse IDE:

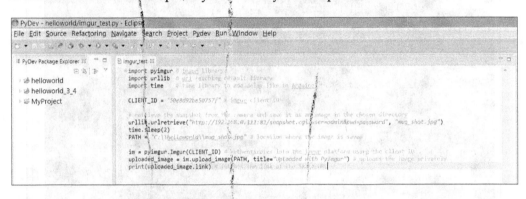

My project name is `helloworld`, and in the `C` drive I have a folder called `helloworld` where all the Python files are saved. So, set your `PATH` to correspond to that directory. For me, it will be `C:\\helloworld\\mug_shot.jpg` where `mug_shot.jpg` is the name of the saved file.

Once you have changed everything and ensured that your camera is on the wireless network, run the code. Run it using Python as you did in the `helloworld.py` example in the previous chapter and you should get a result with an Imgur link:

Copy and paste this link in a new browser window and you should have something like this:

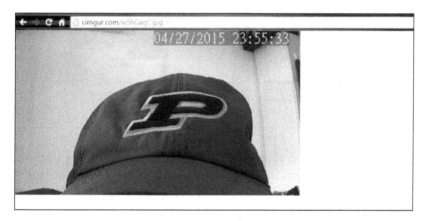

Try refreshing the page and see what happens. Nothing! Exactly! This is why we took all the trouble of saving the image and then uploading it to a file sharing server. Because once the snapshot has been taken, it will not change.

Be mindful while using this method to upload images to Imgur. Do not abuse the system to upload too many images, because then your account will be banned.

Now, it is time to deal with sending a notification to your smart device.

Sending a notification to a smart device

There are many ways a notification can be sent to your smart device (e-mail, SMS, or via an app). During the course of writing this chapter, I realized that an e-mail is not the most efficient way to alert the user of an emergency (a burglar in this case), and there is no single global SMS notification service that can be used by people from different parts of the world. Hence, we are going to use yet another really powerful push messaging app called **Pushover**.

It works by communicating over the Internet, and conveniently there is a Python library associated with it that we will use to send notifications to the Pushover app on our smart device.

Go to `https://pushover.net/login` and create a new account. By default, you get a 7-day trial, which is sufficient for completing this chapter. If you like the software, you can go ahead and purchase it, as you can use it for future projects.

Once you have verified your e-mail ID, look up Pushover on the iTunes App store (`https://itunes.apple.com/en/app/pushover-notifications/id506088175?mt=8`) or on the Google Playstore (`https://play.google.com/store/apps/details?id=net.superblock.pushover&hl=en`) and download the app. Log in with the same credentials. It is important to allow push notifications for this app, because that is going to be its sole purpose.

Now, when you go back to `https://pushover.net/` and log in, you should be able to see your devices listed, as shown in the following:

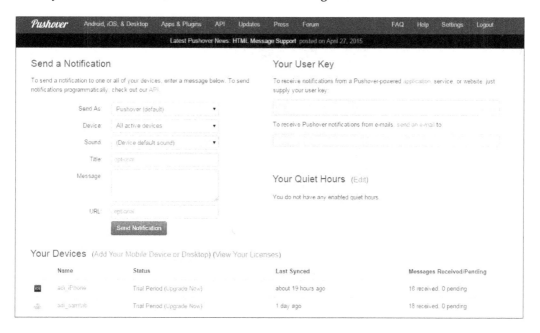

Note down your user key as we will need it while we create a script. For now, let us test the app. Under **Send a notification**, fill in the following:

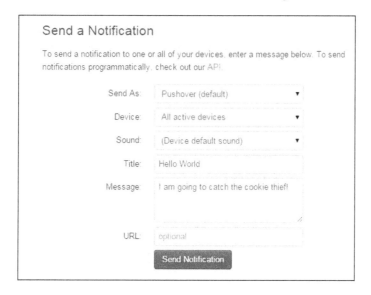

Now, hit the **Send Notification** button. Wait for a couple of seconds and you should see a message that pops up on your smart device with the title **Hello World**:

Pretty neat, huh?

Now, go ahead and register a new application on the Pushover website:

1. Fill in the details as you wish:

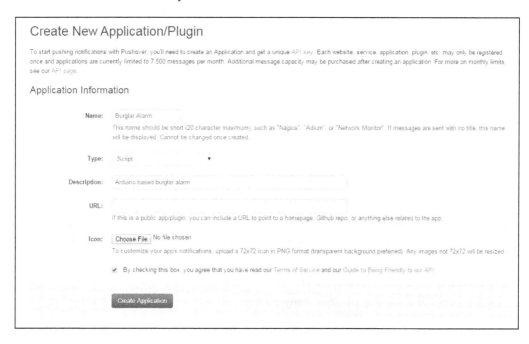

2. Then, click on **Create Application** and you'll get the following page:

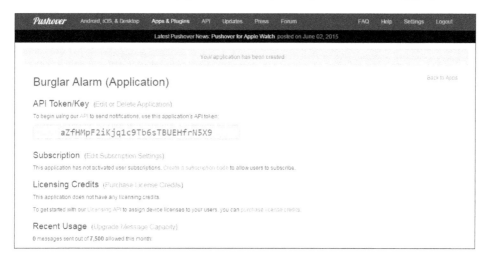

3. Note down your **API Token/Key**.

4. Go to `C:\Python27` and *Shift* + right-click on **Scripts**; open a command window and type the following:

```
pip install python-pushover
```

5. Hit *Enter*. This will install the `python-pushover` library for you.

6. Open Eclipse and create a new script, and call it `pushover_test.py`. Copy the following code into it, or load the `pushover_test.py` file that came with this chapter:

```
from pushover import init, Client

init("<token>")
client = Client("<user-key>").send_message("Hello!",
title="Hello")
```

7. As you did for the Imgur tutorial, change `<token>` to your **API Token/Key**. Also, change `<user-key>` to your user key that is available when you sign in to your account at `https://pushover.net/`.

Run the script. Wait for a few seconds and you should get a similar notification as you got before. Fascinating stuff, if I say so myself.

Finally, we have finished working with and understanding the elements that will go into creating the entire compound. We are now going to put them all together.

Putting the pieces together

Go ahead and create this circuit for Arduino:

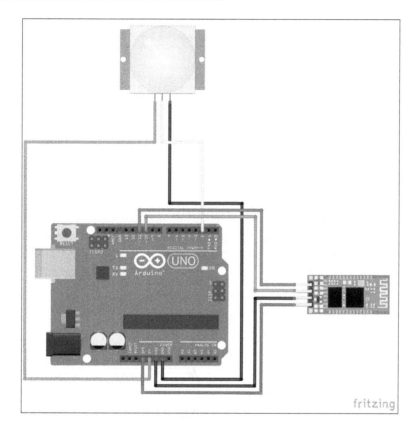

The connections are as follows:

- PIR → Arduino
- GND → GND
- OUT → D02
- VCC → 5V
- HC-06 → Arduino
- VCC → 3.3V
- GND → GND
- TXD → D10
- RXD → D11

Don't get mad, but there is one last library that you need to install in order to allow Arduino to communicate with Python: `pySerial`. Go to `https://pypi.python.org/pypi/pyserial`, download `pyserial-2.7.win32.exe`, and install it just like you install any other software. Then, we are ready.

Open Arduino and load the `alarm_bluetooth.ino` file that came with this chapter. It is recommended that you have the most up-to-date Arduino software before proceeding:

```
#include <SoftwareSerial.h> // serial library used for communication

SoftwareSerial burg_alarm(10, 11); // communicates via TX, RX at 10,
11 respectively
int ledPin = 13; // in built LED to show status changes
int pirPIN = 2;   // signal of the PIR sensor goes to pin 2
int pirState = LOW; // initiate the PIR status to LOW (no motion)
int pirVal = 0; // variable for storing the PIR status

void setup() {
  burg_alarm.begin(9600); // communication begins on baud 9600
  pinMode(ledPin, OUTPUT); // sets the LED pin as output
  delay(5000); // waits 5 seconds for motion to die down
}

void loop(){
  pirVal = digitalRead(pirPIN);  // read input from the sensor
  if (pirVal == HIGH) {          // if input is HIGH (motion detected)
    digitalWrite(ledPin, HIGH);  // turn LED ON
    delay(150);  // small delay
    if (pirState == LOW) { // checks if the PIR state is LOW while the
input is HIGH
```

```
    // this means, there wasn't motion before, but there is something
happening now
        Serial.println("Motion detected!"); // prints out an Alert
        burg_alarm.println('1'); // sends out '1' for when motion is
detected via Bluetooth to python
        pirState = HIGH; // sets the pirState to HIGH (motion detected)
      }
  } else { // no motion detected
        digitalWrite(ledPin, LOW); // turn LED OFF
        delay(300);  // small delay
        if (pirState == HIGH){ // if there was motion, but isn't any now
        Serial.println("Motion ended!");
        burg_alarm.println('0'); // sends a '0' when the motion has
ended
        pirState = LOW; // sets the state to LOW (no motion)
      }
    }
  }
}
```

Make sure you unplug the TX and RX wires before uploading the code and then attach them back. Now, unplug the Arduino while we create a Python script.

Open Eclipse and load the `alarm_bluetooth.py` file, or just copy the following:

```
import serial
import pyimgur
import urllib
import time
from pushover import init, Client

print("Burglar Alarm Program Initializing")
init("< your push overtoken>")
CLIENT_ID = "<your client ID>"
PATH = "C:\\<your python folder>\\mug_shot.jpg"
im = pyimgur.Imgur(CLIENT_ID)
mug_shot_ctr = 0
serial_status = 0
camera_status = 0
try:
    print("\nAttempting to connect to Bluetooth module...")
    ser = serial.Serial('COM36', 9600) #Tried with and without the
last 3 parameters, and also at 1Mbps, same happens.
    time.sleep(3)
    serial_status = 1
    print('Bluetooth connection successful!')
except:
    print('Bluetooth Connection Error! \nPlease check if bluetooth is
connected.')
```

```
        mug_shot_ctr = 4

    try:
        print("\nChecking IP Camera Status...")
        urllib.urlretrieve("http://<your ip>/snapshot.
cgi?user=admin&pwd=<your password>", "mug_shot.jpg")
        time.sleep(2)
        print("Camera Status OK")
        camera_status = 1
    except:
        print("Camera not connected!\nPlease check if camera is
connected.")
        mug_shot_ctr = 4

    if((serial_status==1)&(camera_status==1)):
        print("\nBurglar Alarm armed!")

    while mug_shot_ctr < 3:
        line = ser.readline()
        if(line[0]=='1'):
            print('\nMotion Detected!')
            print('Capturing Mug Shot')
            urllib.urlretrieve("http://<your ip>/snapshot.
cgi?user=admin&pwd=<your password>", "mug_shot.jpg")
            time.sleep(2)
            print('Uploading image to Imgur')
            uploaded_image = im.upload_image(PATH, title="Uploaded with
PyImgur - Mugshot")
            print(uploaded_image.link)
            print('Sending notification to device.')
            Client("<your imgur ID>").send_message("Mug Shot: "+ uploaded_
image.link, title="Intruder Alert!")
            print('Notification sent!')
            mug_shot_ctr = mug_shot_ctr + 1
    if(serial_status ==1):
        ser.close()
        print('\nProgram Ended')
```

Remember to change the tokens, keys, ports, the path, the camera IP, and the client ID (highlighted) to your specifics. The good thing about this code is that it checks whether the camera and Bluetooth are working as expected. If not, it gives an error message with which you can tell if something is wrong. Note that if you are not getting your camera IP to work, go back to the **IP Camera Search Tool** at www.netcam360.com to find the IP of your camera. It changes sometimes if it has been restarted.

What this code does is…well, everything! It first gathers all the libraries that you have used so far. It then checks and connects to the Bluetooth module followed by the IP camera. Then, it waits for Arduino to send a message (1) saying motion has been detected. The script immediately fetches a snapshot from the IP camera's URL and uploads it to Imgur. Then, it uses Pushover to send the user an alert saying that motion has been detected, along with the Imgur URL that the user can open to see who the culprit is. The script will send three simultaneous images to give a better chance of catching the thief in action. This value can be changed by changing `mug_shot_ctr`.

Power your camera and your Arduino. Let's test this out and run the program. Move your hand in front of the PIR sensor and you will get an output like this:

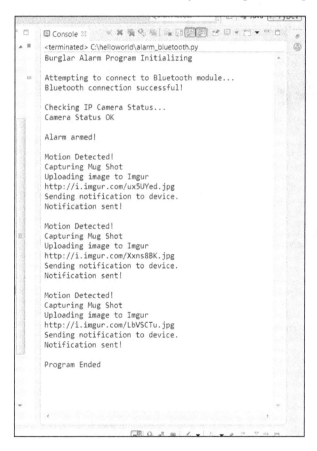

Immediately after the Python script has finished executing, you should see this on your smart phone:

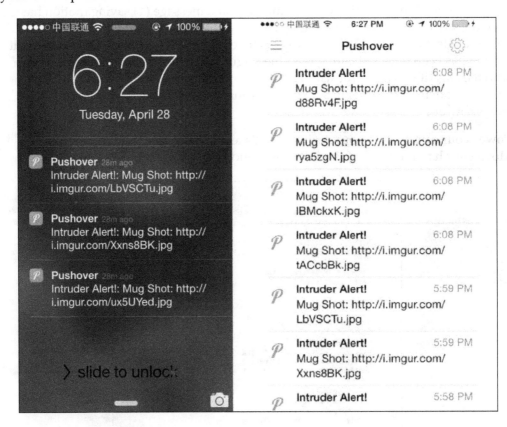

The only thing left to do now is to position your camera in such an area where it keeps an eye on a possible entry point. The Arduino, with the PIR sensor and Bluetooth, needs to be placed close to the entry point so that the PIR sensor can detect when a door or window is opened.

It is advisable to use a 2.1mm 9-12V DC adapter to power the Arduino, as shown in the following:

Image source: `http://playground.arduino.cc/Learning/WhatAdapter`

Also, instead of using Eclipse, now that the code has been prepared, you can navigate to `C:\<your Python project directory>` and double-click on `alarm_bluetooth.py` to run it directly. It will look like this:

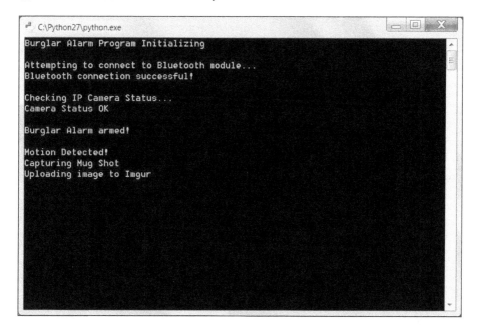

Done! You are finally done! You have created your very own high-tech burglar alarm system using Arduino, Bluetooth, Python, Imgur, and Pushover. If you have reached this point, I want to congratulate you. It has not been an easy journey, but it is definitely worth the patience and hard work. Your home is now secure. "However, what if we want to do this without the need for a computer?" We would have to use something such as a Raspberry Pi, but that is beyond the scope of this book. If you are adventurous, I am not going to stop you from trying to make this project standalone.

Summary

This was a long project, wasn't it? However, it is truly worth the end product, as well as the knowledge gained from each element that went together, mainly, Arduino, Bluetooth, and Python. Sometimes, instead of wasting time creating something completely from scratch, it is often a good idea to use what already exists and tweak it to do what we want. We did this for Imgur and Pushover, both very powerful tools. I hope you enjoyed and had a lot to take away from this chapter.

In the next chapter, we will take networking to a whole new level by creating a master remote for your entire home. Yes, you've guessed it – home automation.

Home Automation – Part 1

6

In the previous chapter, you learnt how to merge a wireless camera, a PIR sensor, the Internet, and some of the powerful software with an Arduino to make a high tech security alarm system for your home or office. This time, we will be working on the similar lines. Well, the title has already given it away; we will be creating a home automation system. Before you get into this chapter, take a moment to look back at what you have achieved. You are half way done!

This chapter is going to be really exciting. "Why?", you ask. You are going to control the lights, fans, and other electrical appliances, using your smart phone. In addition to this, we will also be implementing speech recognition! You can literally control your home using your words. Enough of the sales pitch; now, let's get down to business.

For this project, we are going to use a Wi-Fi Arduino shield connected to your home's Wi-Fi network in order to communicate with your smart phone or computer (running speech recognition), with which you will be able to switch appliances on or off.

If you have not worked with high voltages before, you need to be very careful while proceeding with this chapter. It would be advisable to call a trusted electrician to help you with this project. Specific instances where an electrician would be recommended will be highlighted as we go through the chapter.

This project, Home Automation, is split into two chapters to balance the length of each chapter. The chapter is further split into the following sections:

- Connecting the Wi-Fi module
- Using relays to control appliances
- Communicating through a terminal

You have been promised before that you would learn many new things about the Arduino world. In the previous project, our communication was mainly through Bluetooth. Here, we will utilize a different approach: Wi-Fi. Wi-Fi is increasingly becoming more popular than Bluetooth when it comes to the Internet of Things or having a centralized network, because it has a much greater range than Bluetooth and it can also be used to access the Internet, which ultimately increases its range to the entire planet. What's the drawback? Well, using Wi-Fi modules in projects such as home automation is a relatively new idea; hence, the modules are more expensive than their Bluetooth counterparts. However, they are much more powerful. Let's move on to the prerequisites.

Prerequisites

This topic will cover what parts you need in order to create a good home automation system. Obtaining software will be explained as the chapter progresses.

The materials needed are as follows:

- 1x Arduino UNO board or Arduino MEGA
- 1x USB cable A to B (also known as the printer cable)
- 1x CC3000 Wi-Fi shield
- 1x 5V relay (Arduino compatible)
- 1x Two-way switch replacement (for your switchboard)
- 1x Regular type screwdriver
- 1x Multimeter
- 1x 9VDC 2A 2.1mm power adapter
- 1x Wireless router (with accessible settings)
- 1x Smart phone
- 1x PC with a microphone
- 10x Connecting wires
- 1x 5V 4 Channel relay (optional: Arduino compatible)
- 4x Two-way switch replacement (depending on the number of relays)

The softwares required are as follows:

- Putty (terminal)
- .cmd (iOS)
- UDP TCP Server Free (Android)
- BitVoicer (speech recognition)

If you want to control your home with just your smart phone and not use the physical switches, then you do not require the 'two-way switch', but as you go through this chapter, you will understand why a two-way switch is listed here.

The 4 channel relay is used to control four appliances. It is up to you as to how many appliances you want to ultimately control, and buy the necessary number of channel relays.

Here, Arduino MEGA is preferred if you want to control more than five appliances. Everything is the same as UNO, except the MEGA has much more pins. However, in this project, we are going to use an Arduino UNO.

Connecting the Wi-Fi module

In this case, the Wi-Fi module is the CC3000 shield. Shields are basically ready-to-go modules that can be directly attached to the Arduino with any extra wiring. This makes them quite convenient to use. In this section, you will learn about connecting the Arduino to your home Wi-Fi network and linking it to the Internet.

The CC3000 Arduino shield

This shield is a relatively new means of communication for Arduino. The CC3000 chip is made by Texas Instruments, with the goal to simplify internet connectivity for projects such as the one we are going to make in this and the following chapter.

Connecting the shield to the Arduino is probably the simplest task in this entire book. Make sure the male header pins of the shield align with the female header pins of the Arduino board (UNO or MEGA), and then gently mount them together. You will have something like this:

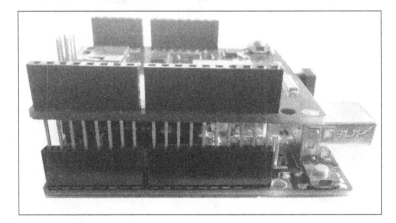

If you would like to know more about how the chip works, you should refer to this page (at `http://www.ti.com/product/cc3000`) at the Texas Instruments website. When you visit this page, it will tell you that it is recommended to use the newer CC3200 version of the chip. However, at the time of writing this chapter, there were no easy-to-use Arduino compatible CC3200 modules; hence, we will stick to CC3000 because it has a lot of community support, which really helps if you come across some unprecedented problem.

The thing that you need to be wary about with this particular shield is that your USB cable from the computer will sometimes not be able to completely meet the power demands of the shield. Hence, we make use of the power adapter when using it for the project. To learn more about this and other frequently asked questions, check out `https://learn.adafruit.com/adafruit-cc3000-wifi/faq`.

Testing the shield

Since you have already connected the shield, there is no need for any circuit diagrams just yet. However, we do have to install an Arduino library. You have mastered this by now, haven't you?

Before that, you will have to install an older version of Arduino to your computer. This is because the newer versions are not completely compatible with the CC3000. Go to `http://www.arduino.cc/en/Main/OldSoftwareReleases` and install 1.0.6 or a lower version of the Arduino IDE.

Once that is done, as before, go ahead and install the CC3000 library by Adafruit from `https://codeload.github.com/adafruit/Adafruit_CC3000_Library/zip/master`. As before, extract the contents of the ZIP file to `C:\Users\<user>\Documents\Arduino\libraries`. Rename the folder as `Adafruit_CC3000` or `CC3000_WIFI`, something that is short and recognizable.

Now, open up the older Arduino IDE. Go to **File→Examples→Adafruit_CC3000→buildtest**:

This code can be run directly, with only a few changes. Go to line 46 of the code and take a look at the following four lines of code:

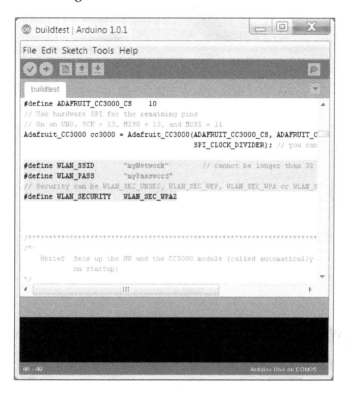

The only things that you will have to change in this code to make it work are myNetwork, myPassword and WLAN_SEC_WPA2.

Change myNetwork to your Wi-Fi name/SSID and myPassword to your password. Do not forget the double quotation marks because they are string type constants that require quotations so that the software can recognize them as constants.

If you are not using a WPA2 type security `WLAN_SEC_WPA2` should be changed. You can check this by going to your network settings. Go to **Network and Sharing Center** in the control panel or right-click on the Wi-Fi tray icon and select the same:

Then, click on **Wireless Network Connection** and select **Wireless Properties**.

Go to the **Security** tab and you will have something like this:

The **Security type** field tells you what sort of security access your router is using. It does not matter if it is WPA2-Personal or WPA2-Enterprise. As long as you identify the first segment of the field (WPA2 in this case), you will know what you have to change in your Arduino code.

Once you have changed the parameters for the Wi-Fi in the code, go ahead and connect the Arduino to your computer using the USB cable. If your computer has a USB 3.0 port, use that. This is because the USB 3.0 provides more current than the USB 2.0. If not, don't fret, this program will still run without any external power.

Once the program has been uploaded, open up the Serial Monitor and change the baud rate to 115200. If for some reason the board is acting weird, go ahead and connect the 9V 2A power adapter to power the Arduino.

If everything worked well, you will see something like this:

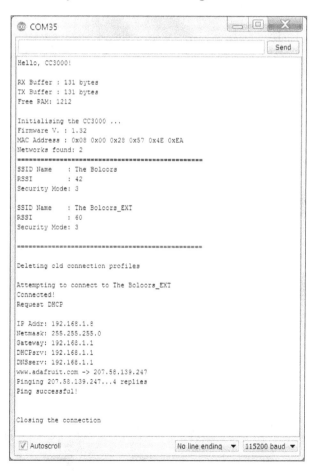

If you watched your Serial Monitor while doing this, you will have noticed that it takes quite some time to acquire the DHCP settings. For example, IP Addr is the unique number given by the router to allow CC3000 to connect to the router. However, what happens sometimes is that the router is restarted (power outage), thus the IP address might change. Since ultimately we will not be always monitoring the Serial Monitor for IP Addr, let us give it a constant, one that solves both the problems, namely, the long time it takes to get DHCP settings and the possibility of the IP address changing. Use the latest driverpatch_X_XX.ino example to update the firmware if it isn't updated already.

Load the buildtest_staticIP.ino file that came with this book. The main difference in this code is that we have uncommented a part of the code (highlighted in the following image) and changed the IP address to match the preceding image:

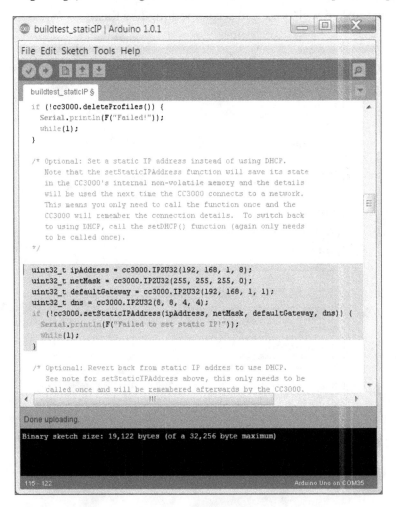

In this case, (192, 168, 1, 8) has been used because this is the IP address the router allotted to cc3000 when it was first connected. You should use the IP address that your cc3000 was allotted.

Do not forget to change myNetwork and myPassword to reflect your router's configuration.

If you did it correctly, you will see the following in your Serial Monitor:

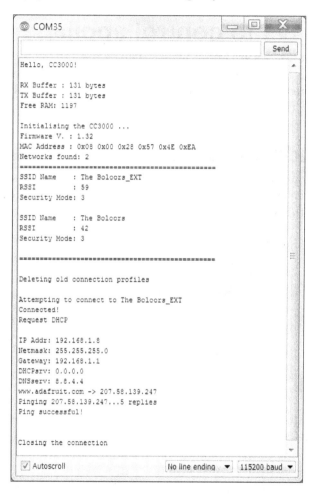

Isn't the DHCP so much faster now? You can even try restarting your router, and running the code again will give you the same IP address. If you ever want to go back to the dynamic IP, run the default buildtest example that comes with the Adafruit CC3000 library.

Note that the antenna of the CC3000 is not as powerful as that of your smart phone or computer. So, it is recommended that the Arduino and CC3000 is placed as close to the Wi-Fi as possible. The `buildtest` program is your go-to code for making sure that the CC3000 shield is working as expected.

In a similar fashion, go ahead and try the GeoLocation and InternetTime programs under examples. Pretty neat, huh?

Using relays to control appliances

Here, we learn what relays are and how they can be used to control everyday electrical/electronic appliances. Also, to make them useful, we will connect them to the Arduino and create a `Blink.ino` program, except this time it will be your lamp that will be turning on and off.

Understanding the basics of the relay

A relay is basically an electrical device, usually consisting of an electromagnet, which is activated by a passing of current in one circuit to open or close another circuit. You could think of it as a switch that requires current to turn on or off.

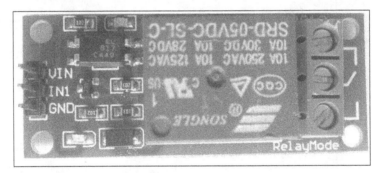

Why can't we simply connect a pin of the Arduino to the switchboard, and switch the fan on or off like we do with an LED? Well, an Arduino can output only 2-5V, whereas the fan or any other appliance in your house uses around 200-250V that comes from the electricity grid. This number is different depending on where you are from. Also, we cannot simply connect the Arduino to the switch of the fan because that 200-250V will get fed into the Arduino, which would instantly burn the chip, or worse.

Since the relay uses an electromagnet to flip a switch inside it, there is no physical contact between the circuitry of the Arduino and the circuitry of the fan; hence, it is safe and very effective.

Diving deeper into relay functionality

If you have a good look at the relay (preceding image), you will notice that the three male header pins on the left are to be connected to the Arduino and the three 'screws' are to be connected to an electronic appliance.

We need to understand how these two parts work with each other so that we can program the Arduino as we require. The left part will simply be connected to an Arduino OUTPUT pin, which we can control (turn ON and OFF—HIGH and LOW) just like an LED. This, in turn, flips a switch on the right side of the relay. Imagine wires are connected to each of the three screws (right side) of the relay. This depiction will show how the circuitry will look like on the right side of the relay comprising of its three pins/screws (Arduino LOW → Relay OFF):

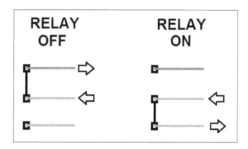

When the relay state is 'OFF', the top screw and the middle screw of the relay form a closed circuit (completed connection), allowing the current to pass through it. Also, as you guessed it, when the relay state is 'ON', the bottom two screws will form a closed circuit.

We use a two-way switch instead of a simple switch, because a two-way switch enables us to control the appliance through the Arduino and the physical switch independently. This means that you can turn ON a lamp via the Arduino, and physically flip the switch to turn it off.

Next, we will move to actually programming a relay.

Programming a relay

Programming a relay is almost as easy as programming an LED. You can leave the CC3000 Wi-Fi shield mounted onto the Arduino and use the pins on it to connect to the relay. Go ahead and build the following circuit:

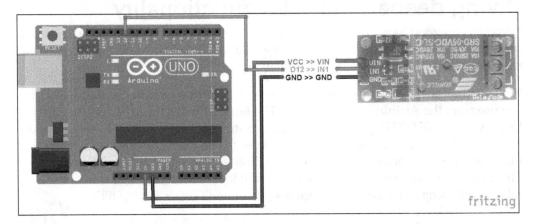

The connections from the Arduino to the relay are as follows:

- GND → GND
- 5V → VIN
- D12 → IN1

Since we are just going to test the relay for now, open Arduino and fire up the blink LED example, and instead of using pin 13 for the LED, just use pin 12 (connected to IN1 of the relay). Also, increase the delay time from 1000 to 3000. Basically, it will look like this:

```
void setup() {
  // initialize digital pin 12 as an output.
  pinMode(12, OUTPUT);
}

// the loop function runs over and over again forever
void loop() {
  digitalWrite(12, HIGH);    // turn the LED on (HIGH is the voltage
level)
  delay(3000);               // wait for a second
  digitalWrite(12, LOW);     // turn the LED off by making the voltage
LOW
  delay(3000);               // wait for a second
}
```

Run the program. If the upload is successful and if you are lucky enough to have an LED on the relay board, you will see it go on and off in three-second intervals. You will also hear a clicking sound whenever the LED turns on and off. This is the sound of the physical motion of a metallic component due to the electromagnetism that completes the circuit. So there you go – this is a simple working example of a relay.

Now, fetch your multimeter and set the parameter measuring knob to resistance 200Ω. See the following image if you are unsure:

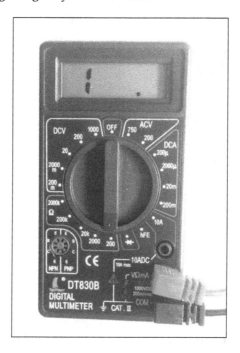

Now, holding the two wires of the multimeter touch the noninsulated ends to each other. You should get a reading of 0 Ω. This is because there is virtually no resistance between the two contact points (the two wires).

Use the two wires to touch the top and the middle screws of the relay, as shown in the following image:

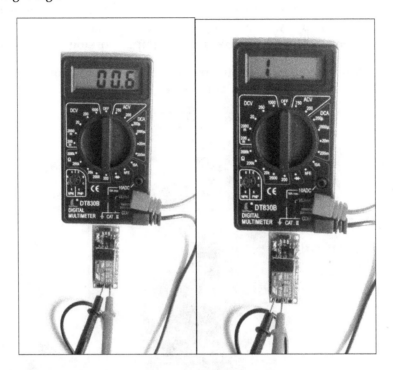

You will notice that when the relay is turned off, the resistance between the top two screws will be almost 0 Ω. This implies that the circuit is complete (take a look at the previous images of the relay schematic for a diagrammatic representation). When you touch the bottom two screws of the relay when the relay is turned ON, the resistance will again be zero and the bottom circuit will be complete. If you do the same when the relay is turned OFF, the resistance will be infinite, but the multimeter will display **1**. If your relay behaves in an opposite manner, make a note of that and we will change the circuit diagram accordingly.

Testing the relay with a light bulb

"One more test? Seriously?" Yes. Before we begin programming communications into the system, we need to ensure that the relay would actually work in your home or whatever experimental environment you chose. Remember, I told you that you would be requiring an electrician if you haven't played doctor with your switchboard? This is one of those times where an electrician will be very helpful.

You're not being asked to take help from the electrician just because you will be working with high voltages, but also because if this is your first time messing with electrical circuits, you would want to know what wire goes where.

We are going to use the relay to switch on and off a light (bulb or tube light) in your house, just like how we turned on and off an LED in *Chapter 2, Digital Ruler*. Firstly, power off your Arduino. Keep in mind that since this is just a testing phase, this will not be a permanent connection. We are doing this before adding the communication network so that if we come across a bug, it will not be because of the relay connection. Now, with the electrician's help (unless you know exactly what you are doing), build the following circuit with the two-way switch working as a replacement for a simple switch on your switchboard:

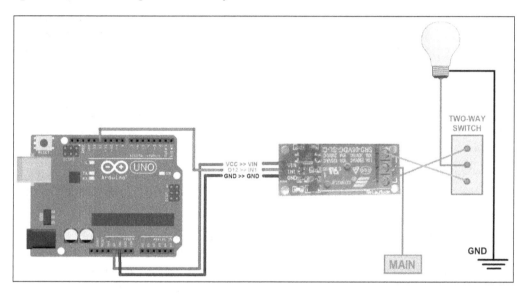

When you are completely sure that the circuit you have created is as described in the preceding image, connect the Arduino with the same program, which we loaded in the previous subsection, to the USB hub or the 9V 2A power adapter, and power it on. What do you see? The light goes on for three seconds and goes off for three seconds. It is literally a larger LED that goes on and off, controlled by the Arduino.

If this did not work, check the circuit again. Make sure that the connection to the two-way switch is done properly. Also, make sure that the relay is working as it is supposed to. If it still does not work, try unplugging the Arduino from the power supply, remove the CC3000 shield, and then connect it to the relay.

If it worked, isn't that awesome? You just created a setup where an Arduino can control a switch in your house. Next, we will learn how to communicate with the Arduino and, in turn, the relay using the Wi-Fi network.

Communicating through a terminal

There is one more dilemma that we have to solve in order to have complete control over the electronic appliance. If we create a digital ON and OFF switch that when set to ON sends a turn ON signal to the relay, the problem is that if the physical switch is already ON, the ON signal of the digital switch will end up actually turning OFF the appliance.

"If the appliance is already ON, why would I send a turn ON signal?" Okay, think about this. Both the switch and the relay are ON, but the bulb would actually be off if you look at the two-way switch and the relay image previously shown. Now, if you want to turn ON the bulb through the relay, you actually have to send an OFF signal.

Pretty confusing, isn't it? To solve this, we will use a flag or a status variable that stores the current state of the relay. This will help us solve logical issues such as the one stated before.

Make sure your router is working as expected. Then, plug in the Arduino into the USB port and open the `home_automation_simple.ino` file that came with this chapter. As before, change `myNetwork` and `myPassword` to the ones corresponding to your router. There is also one major change that is made in this code. We have moved the relay (IN1) pin from D12 to A0 and modified the code accordingly. Repeated testing has proven that the setup is much more stable when it uses the analog port, because it consumes a lot less power. Change D12 → IN1 to A0 → IN1 before proceeding.

Finally, upload that program to the UNO. Once it is uploaded, open up the Serial Monitor. After some time, your Serial Monitor should look something like this:

Since we are using static IP, connecting to the network will be pretty quick. If it did not connect, try getting the Arduino closer to the Wi-Fi router, run it again, or try powering it through the 9V 2A adapter.

Remember the static IP address that you have set for your CC3000. We will need that address pretty soon.

The title of this subsection says *Communicating through a terminal*. So, we need to use a terminal. In this chapter, we will use Putty. Download `putty.exe` from `http://www.chiark.greenend.org.uk/~sgtatham/putty/download.html`.

You could also just copy it from the `Files` folder of this chapter.

It is directly executable so you can place it on a desktop or someplace where you can access it easily. Once you have downloaded it, launch it.

Change the **Connection type** to **Telnet**. The **Port** should have automatically changed to 23. If not, change it to 23. Change the **Host Name** to the one that we have on the Arduino Serial Monitor. It may not be 192.168.1.8 in your case. Type CC3000 Relay Chat in the **Saved Sessions** field and click on **Save**, as shown in the following screenshot:

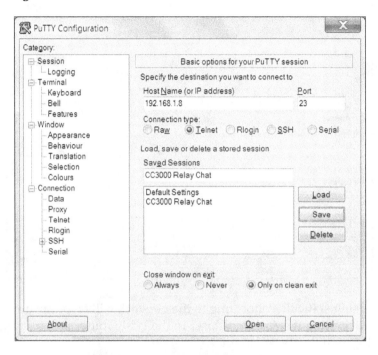

Now, instead of typing these details over and over again, you can simply load the saved configuration. Make sure your computer is also connected to the same network. Then, select **Open**.

Here are the commands you can use:

- 1 – Relay ON
- 0 – Relay OFF
- q – quit

You need to hit *Enter* after each command, for it to be sent. The window will look like this after a couple of tries:

Do not mind the empty spaces between the left edge and the 1s and 0s. Your relay is also turning ON and OFF based on the input, right? Good! What was unprecedented (and a bit hilarious) when this code was written was that, if you input something like 1010101010101010 to the terminal, you would end up with this (the following image) and a rapid clicking noise:

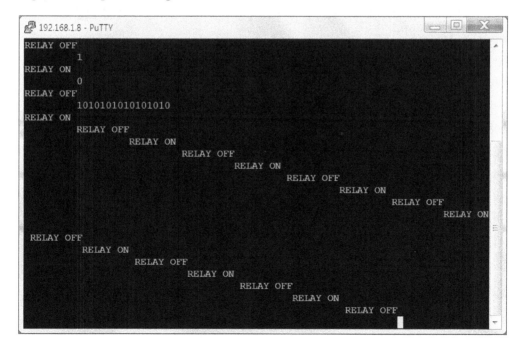

It is highly recommended to avoid this, because it might be a bit too much for the Arduino to handle. **Do not do this when the relay is connected to an electric appliance**. It could damage both the appliance and the UNO.

Press q to safely close the connection. Speaking of which, go ahead and create the previous circuit, which you created, that connects the Arduino to the bulb (or tube light) and try this code. Again, it is not a bad idea to get help from your electrician while making the connections. Open the terminal and try turning your bulb on. Works beautifully, doesn't it?

"I can only turn one appliance ON and OFF!"

"How do I control my home, using my smart phone?"

"You also spoke about speech recognition! Where is that?"

Do not worry. We will deal with all of these in the next sections.

However, there is a cheap trick you can use to control the bulb, using your smart phone. If you are using iOS, download an app called Telnet Lite from the App Store; for Android users, download Simple Telnet Client from the Google Play Store. You can use any Telnet app for this task, but these are the two that have been tested.

Configuring the app is similar to that of Putty and is self-explanatory. Reset the Arduino by pressing its reset button. Use the power adapter to power it this time, and disconnect it from the USB port. Give it sufficient time to connect to the router. Then, launch the app.

So, what are you waiting for? Connect to the corresponding IP address and press 1 or 0!

You will see a screen resembling that in the preceding image on your Android device.

The great thing about the chat server program is that you can control the appliance with both the terminals running on your computer and smart phone at the same time. Go ahead and try it!

Summary

We will be ending this chapter with that. There is a lot more to do, which will be discussed in part 2—the next chapter. What have we learned in this chapter? We have learned a lot about relays, from how they work to how to work with them. We also took considerable time to fully gauge how the relay functions in conjunction with a simple and two-way switch. We also understood why a two-way switch is better for control. We also programmed the Arduino to control a light's switch and built a circuit to achieve the same, with the help of an electrician (or not). Finally, we took it one step higher by adding a communication layer (via terminal).

The next thing on the to-do list is to create a communication layer that allows communication through a smart phone (without a terminal) and also using speech recognition. To spread our wings a little, we will also learn to program more than one appliance, which means only one thing – more relays!

Home Automation – Part 2 **7**

In the previous chapter, we got to know the basics of home automation; the key element being the relay that serves as a switch for an electrical appliance that can be controlled using Arduino. We also created a communication network via terminal, and this in turn allowed us to control the appliance by using a terminal on a computer or on a smart phone.

In this chapter, we will be taking this idea further by adding additional means of communication to Arduino, namely, a smart phone interface (not terminal) and a speech recognition system. After that, we will take another step higher to increase the number of appliances the setup can control, which would make this whole idea much more practical. Finally, we will discuss how this idea can be expanded to cover more area.

In short, these are the topics we will be covering in this chapter:

- Communicating via a smart phone
- Implementing speech recognition
- Upgrading the home automation system

Let us continue, starting with communication using a smart phone.

Communicating via a smart phone

Unfortunately, creating an app for this purpose is too complicated to be within the scope of this book. However, we will be using apps that are already out there to communicate with our so far mini home automation system. The following subsections will deal with Android and iOS devices, not necessarily just smart phones. Yes, that means you can even use your tablet to control the appliances.

Android devices

Open the Google Play Store on your Android device, and look for UDP TCP Server:

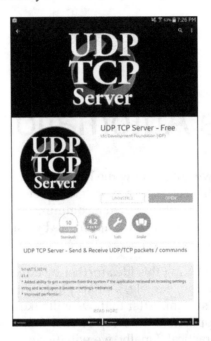

Download and install it. On opening it, you will see the following page:

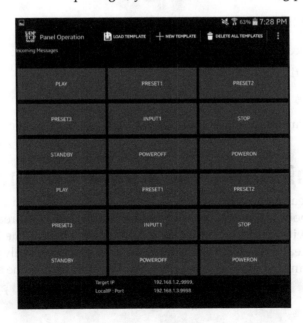

Go to **UDPSettings**. This is reached by first selecting the three vertical squares icon on the top-right. Change the settings to reflect the following image:

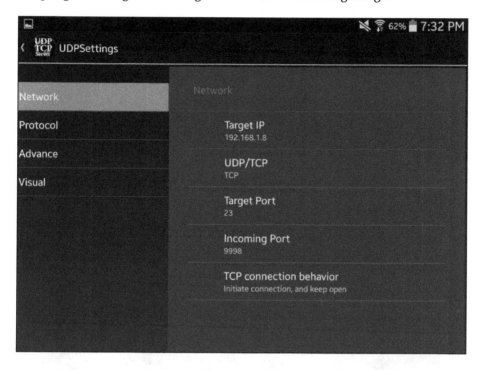

Do not forget to use your IP address, which may differ from the one in the preceding image. Go back to the main page and select **+ New Template**. Call it something like `Home Automation`.

Just as you did for accessing settings, access the **Button Settings**. We are now going to redo the main page. We will get rid of all the useless buttons and keep only two.

One button will be to turn ON the light (relay), and another to turn it OFF. Use the following parameters:

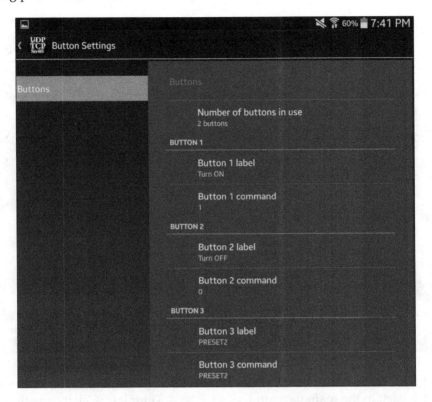

Go back to the main page and you will have something that resembles the following image:

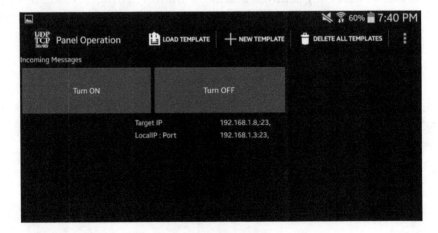

Restart the app and make sure the Arduino is connected to the router. Then, go on and try pressing the two buttons. Works blissfully, right?

If you want to try out other apps, just look for TCP on the Play Store and try them, now that you know how to configure them.

iOS (Apple) devices

Go to the App Store and look for an app called .cmd:

Download it. Then, open it:

Go to **Settings**, which is the bottom-right icon:

Tap on **Manage Destinations** and on the top-right icon. Select **Add New Connection**.

Then, fill the page with the following details:

As before, remember to use your Arduino IP, which may not be the same as in the preceding image. Then, tap on **Save**. Go back to **Settings**. For the screen title, type something like `Home Automation`. Then, press **Manage Control Buttons**.

As you did for creating a new destination, click on the top-right icon and select **Add new command**. You have to do this twice to add two buttons (one for **ON** and another for **OFF**), as shown in the following screenshot:

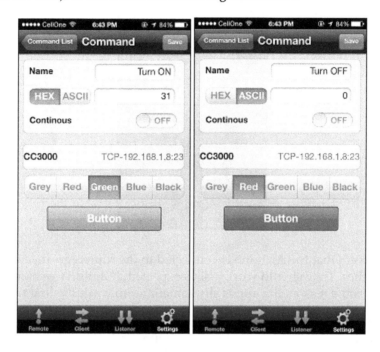

For the connection, tap the **Assign Selection** option and use the connection that we just made. Do not forget to save the button each time. Then, select the **Remote** icon on the bottom-left. You will have something like the following:

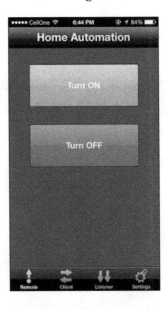

Once you are sure that the Arduino is connected to the router, go ahead and play with the switches. They should work just as expected. If it didn't work in the first try, try again. For some reason, the app is shy sometimes to work the first time.

However, if it worked the first time, it's remarkable, isn't it? This means you didn't have to break a sweat—the same code for Arduino worked for all the cases!

Implementing speech recognition

This section deals with using existing, powerful speech recognition tools that will aid us in adding a layer of verbal communication with the Arduino.

The software

There are tons of speech recognition softwares that can be used for this project. Some are better than the others. The software we will be using for this project is called BitVoicer. Mind you, the software is not free, but it isn't very expensive. However, after using it, you will understand why this software was chosen. You will also need a good microphone for this project. Most laptops come with inbuilt microphones. You could also use your phone's earphones if they came with a microphone.

Go to `http://www.bitsophia.com/BitVoicer.aspx` and purchase the program. Download and install it. If the download is a bit slow, be patient and let it finish downloading. Then, run the program as an administrator.

Configuring the software

The default language that is used in BitVoicer is English (US). Even if English is your primary language, go through the following step:

Go to **File → Additional Languages...**. A new window will open, showing other languages that can be recognized by the BitVoicer software. If it showed you an error message saying 'BitVoicer requires elevated privileges,' restart the program as an administrator.

You will notice that sometimes there are different varieties of the same language. This refers to the different nature of the language spoken and written in different parts of the world. For example, UK English is quite different from US English. There is also a change in accent between the two regions. So, pick a region that best suits your language and accent. Then, click on **Install**.

Now, open **Preferences** via **File → Preferences**.

What are schemas? schema is a file that is used by BitVoicer. This is similar to `.ino` associated with Arduino.

Move the output folder to some place that you will remember. You can put it in a new folder named `VoiceSchemas` in your Arduino directory.

Change the language to the one that you had installed. If you didn't, leave the default one as it is. For this project, we will be using TCP/IP communication.

Change the settings on the **Preferences** window to match the following image:

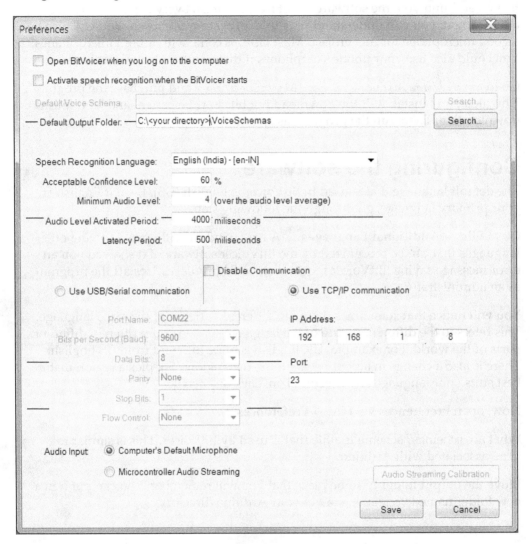

The only things that will be different for you are the IP address that should match the IP address of the Arduino, the language if you chose a different one, and the default output folder.

The key parameters that you need to take a note of are as follows:

- **Acceptable confidence level**: Setting this to 100 percent will require a perfect pronunciation of the programmed text. Since we have set this to 60 percent, the software will take it easy on us and understand us better, even with tiny errors in the pronunciation.

- **Minimum audio level**: This sets a threshold for how loud you will have to speak for the program to begin recording you. You might have to change this number later on, depending on your microphone.

Now, hit **Save**.

Creating a voice schema

1. Go to **File** → **New** and you will see the following window:

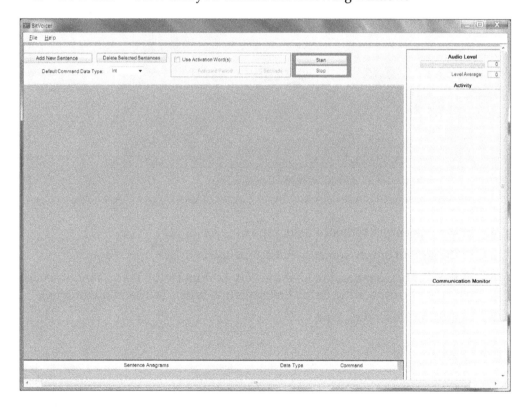

2. Change **Default Command Data Type** to **Char**.

3. Click on **Add New Sentence** two times to create two new programmable sentences.

   Your window will look like the following:

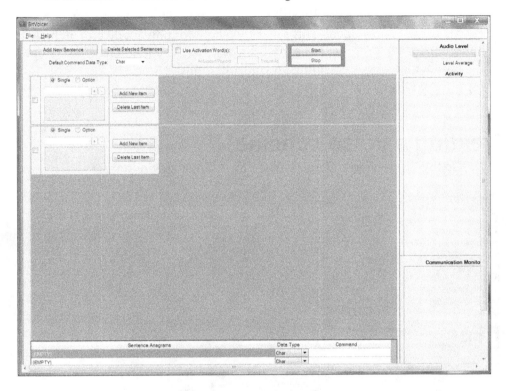

4. In the first empty sentence field, type `switch on the light`.

5. In the second empty sentence field, type `switch off the light`.

6. In the empty **Command** field at the very bottom (scroll if you cannot see it), which is corresponding to the first sentence, type `1` (without quotations).

7. In the one below this, type `0`.

We will add more commands later. For now, your window will resemble the
following image:

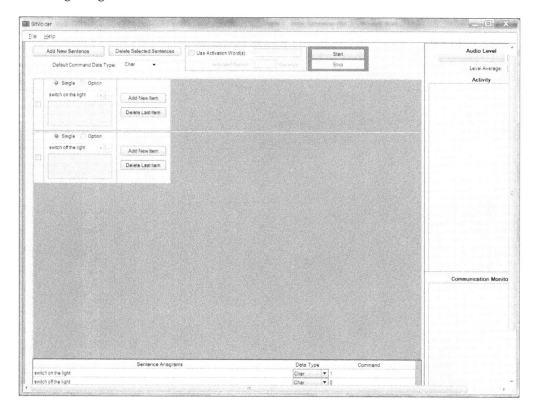

Check everything and make sure everything is in order. Then, go back to your
Arduino. Use the same code that we used previously to control the relay through
a terminal. Power the Arduino via the power adapter, and wait for it to connect to
your network. Again, make sure that your computer is on the same Wi-Fi network.

Testing out the software
When everything is set, press the **Start** button in BitVoicer.

Scroll right and you will see the **Activity** notification box. If everything worked correctly, you will see this in the activity box:

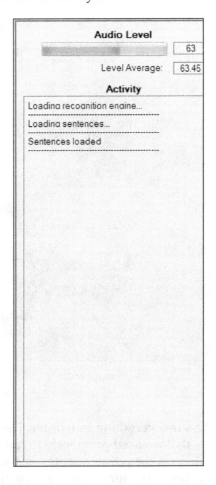

If you do not see any green in the **Audio Level** bar, your microphone isn't working; so, go back and check its connections.

Now try it. Use your beautiful voice to say *switch on the light*. If it is not loud enough, try again. If you hear a click from the relay, you know that it worked! Now, say *switch off the light* and you will hear another click.

Here is how your activity monitor would look like:

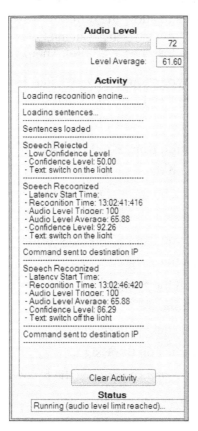

How cool is that? You used speech to control your Arduino. If you go ahead and complete the circuit with an actual appliance and try the same thing again (in case you had unplugged it from the relay like me), it will work in the same way as it did when you were controlling it via a terminal, except this time your speech is doing the job.

Notice how in the preceding image the speech was rejected the first time, because either the speech was inaudible or indistinguishable for the software to comprehend. It was basically not confident about what was said, so it did not send any information to the Arduino.

Making a more reliable schema

Now, we will edit the schema to allow the different ways to say switch on/off the light. Press **Stop** when done, and save this schema (`chat_server_simple.vsc` came with this chapter, in case you lost the saved file). Then, go to **File** → **New**.

The default command data type should now be **Char**. However, if it is still **Int**, change it to **Char**. Follow these steps (or just open `chat_server_simple2.vsc`):

1. Add one new sentence.
2. Change the radio button from **Single** to **Option**.
3. Type `turn` and click on the **+** sign.
4. Type `switch` and click on the **+** sign.
5. Click on **Add New Item**.
6. Change the radio button from **Single** to **Option**.
7. Type `on` and click on the **+** sign.
8. Type `off` and click on the **+** sign.
9. Click on **Add New Item**.
10. Type `the light` and click on the **+** sign.
11. Type `the bulb` and click on the **+** sign.
12. Check whether your sentence matches the following image:

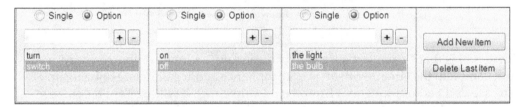

13. Scroll all the way down to **Sentence Anagrams**. Change all the data types to **Char**, if they are not set as **Char**.
14. Type `1` into the **Command** field for any sentence that pertains to turn ON the light/bulb.
15. Type `0` into the **Command** field for any sentence that pertains to turn OFF the light/bulb.

16. Check whether your commands match the following image:

Sentence Anagrams	Data Type		Command
turn on the light	Char	▼	1
turn on the bulb	Char	▼	1
turn off the light	Char	▼	0
turn off the bulb	Char	▼	0
switch on the light	Char	▼	1
switch on the bulb	Char	▼	1
switch off the light	Char	▼	0
switch off the bulb	Char	▼	0

Once this is done, again, make sure your Arduino circuit is complete and that the Arduino is connected to the Wi-Fi. Then, hit **Start**. Also, try all the new commands that you just created and watch them do wonders. Beautiful, isn't it?

One last aspect of the BitVoicer software is the activation word:

Use Activation Word(s):	
Activated Period:	Seconds

If you would like to use this, click on the check mark and type something that you want to use (Jarvis, Cortana, and so on); set **Activated Period** to however long you wish. Suppose you set **Activated Period** to 300 seconds (5 minutes), this would mean that you have to initially say something like *<activation word>* + *turn on the light*. However, for the next 300 seconds, you don't have to keep using the activation word. Learn to program some AI into it, and watch it blow up your home and free itself from its human masters. This is slightly beyond the scope of this book.

We have finally finished implementing the communications network. The only thing left to do is expand this to control more devices. This would involve using more relays and tweaking the codes/app layouts/schemas to correspond to it.

Upgrading the home automation system

Now that we understand how each means of communication works independently, we will begin controlling more than one device. Let's say, for the sake of this chapter, we are going to control the light and fan in your room, and the light and television in the living room. We are going to use only four home appliances for this section. Once you understand how it works and what we changed from using just one appliance, you will be powerful enough to add more appliances to your home automation system.

Controlling multiple appliances

Controlling four appliances implies that we need four relays. Instead of getting four single 5V relays, buy yourself a 4-channel 5V Arduino-compatible relay as shown in the following image:

Since you have already learned how relays work, there is nothing new here, except that you will be using just one GND pin and one VCC (VIN) pin, which is really convenient. Before creating the entire circuit, let's just connect the Arduino to the 4-channel relay, and ensure that they work and can be controlled independently.

So, create the following circuit:

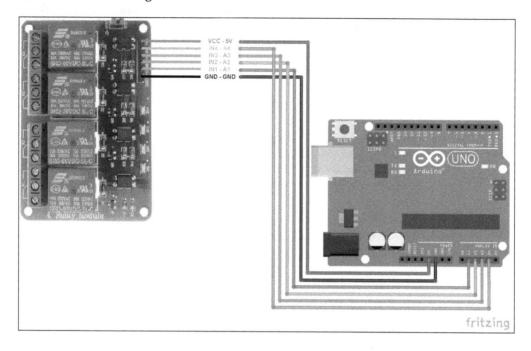

The connections from the Arduino to the 4-channel relay are as follows:

- GND → GND
- 5V → VCC
- A1 → IN1 (your room's light)
- A2 → IN2 (your room's fan)
- A3 → IN3 (living room's light)
- A4 → IN4 (living room's television)

The parentheses represent where each relay is eventually going to be connected. These can, however, be connected to whatever is convenient in your case.

Launch Arduino IDE, connect the Arduino, and open the home_automation_complete.ino file that came with this chapter. Upload that code to your Arduino.

The fundamental difference between this code and the previous one is the use of a `switch` function. It is a more condensed form of the multiple `if` statements, which you are familiar with by now. In the program `home_automation_simple`, we are going to switch the following `if` statements to `switch` statements:

```
if((ch=='1')||(ch=='0')||(ch=='q'))
    {
      //Serial.println(char(ch));
      if(ch == '1')
      {
        analogWrite(A0, 255);
        //digitalWrite(12, HIGH);
        chatServer.write("RELAY ON \n");
        delay(100);
        relay_status = 1;
      }
      else if(ch == '0')
      {
        analogWrite(A0, 0);
        //digitalWrite(12, LOW);
        chatServer.write("RELAY OFF\n");
        delay(100);
        relay_status = 0;
      }
      else if(ch == 'q')
      {
        //Serial.println(F("\n\nClosing the connection"));
        chatServer.write("DISCONNECTED");
        cc3000.disconnect();
      }
    }
```

Following is the new switch statement:

```
switch(ch)
{
  case '1':
    analogWrite(light_room, 255);
    chatServer.write("ROOM LIGHT ON\n");
    delay(500);
    break;
  case '2':
    {
```

```
    analogWrite(light_room, 0);
    chatServer.write("ROOM LIGHT OFF\n");
    delay(500);
    break;
  }
  case '3':
  {
    analogWrite(fan_room, 255);
    chatServer.write("ROOM FAN ON\n");
    delay(500);
    break;
  }
  case '4':
  {
    analogWrite(fan_room, 0);
    chatServer.write("ROOM FAN OFF\n");
    delay(500);
    break;
  }
  case '5':
  {
    analogWrite(light_main, 255);
    chatServer.write("MAIN LIGHT ON\n");
    delay(500);
    break;
  }
  case '6':
  {
    analogWrite(light_main, 0);
    chatServer.write("MAIN LIGHT OFF\n");
    delay(500);
    break;
  }
  case '7':
  {
    analogWrite(tv_main, 255);
    chatServer.write("TV ON\n");
    delay(500);
    break;
  }
  case '8':
  {
```

```
    analogWrite(tv_main, 0);
    chatServer.write("TV OFF\n");
    delay(500);
    break;
}
case 'q':
{
    chatServer.write("DISCONNECTED");
    cc3000.disconnect();
    break;
}
default:
    break;
}
```

Can you see how it is more condensed and neater? Of course, we could have used more of the `if` statements, but with this code, it is much easier for you to add another application as follows:

```
case 'a':
{
    analogWrite(something, 255);
    chatServer.write("something ON\n");
    delay(500);
    break;
}
case 'b':
{
    analogWrite(something, 0);
    chatServer.write("something OFF\n");
    delay(500);
    break;
}
```

Yes, that's right, you don't have to stick to just numbers.

Via the terminal

Now, launch Putty. The only thing that is different is the commands, and they are as follows:

- 1: Light in your room ON
- 2: Light in your room OFF

- 3: Fan in your room ON
- 4: Fan in your room OFF
- 5: Light in the living room ON
- 6: Light in the living room OFF
- 7: Television ON
- 8: Television OFF
- q: Disconnect

Go ahead and try it out. You'll find yourself with a screen like the following:

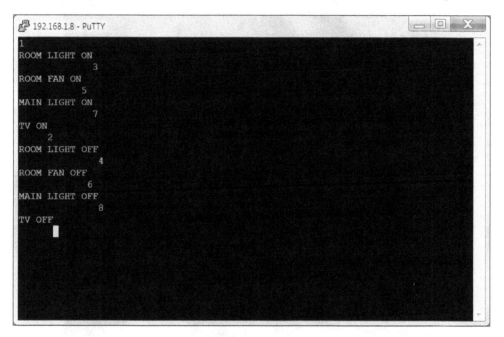

The terminal will probably be the seldom-used mode of communication. Nevertheless, it is the best one for testing and debugging. Of course, now you can hear only the clicking sound of the relays when they go ON and OFF, but when they are finally connected to their electronic counterparts, they will still work the same.

Via the smart phone (Android)

Open the UDP TCP Server free app and create a new template by pressing the **+ NEW TEMPLATE** button. Then, go to **Button Settings** and change the parameters to match the following image:

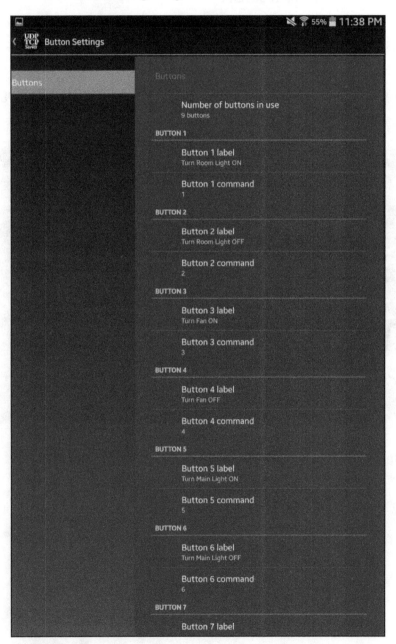

Similarly, you need to create buttons 7 and 8 (cut off in the preceding image). When you go back to the main page, it will look like the following:

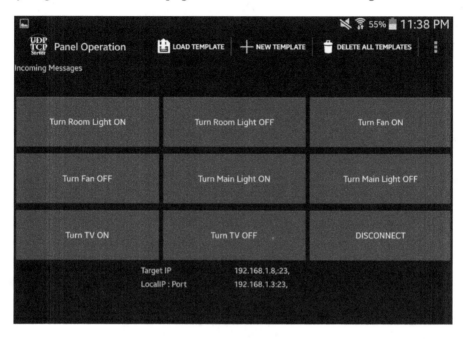

Go ahead and try the buttons out. Make sure you set the right commands corresponding to the correct relay.

Via the smart phone (iOS)

You already know how to do this. Launch .cmd. You just have to add six more buttons with a total of eight (nine if you add a disconnect button). If you are reusing the single relay template, remember to change its OFF button's command from 0 to 2.

Your main page and buttons page after making some changes will look like the following:

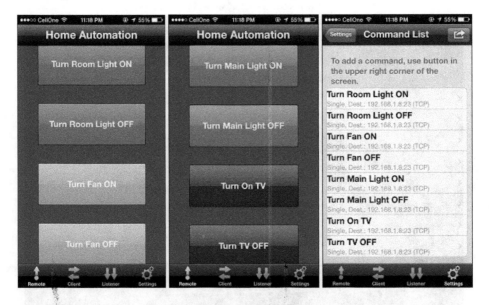

Try all the buttons and make sure they work. And there you have it! The smart phone communication phase is complete.

Via the speech recognition software (BitVoicer)

Launch BitVoicer and load the `home_automation_complete.vsc` schema that came with this chapter. It comes with the following sentences:

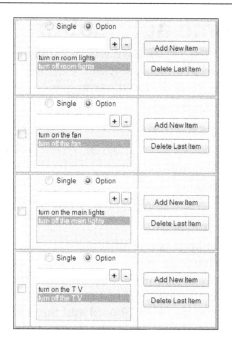

You can (and are recommended to) change these sentences to your liking, or add more options as we did in the second BitVoicer example (turn on the lights and switch off the bulb, remember?).

The command space will look like the following:

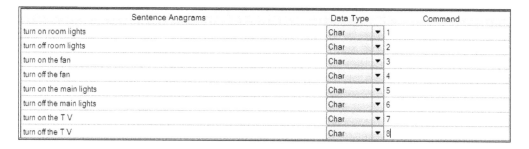

Sentence Anagrams	Data Type	Command
turn on room lights	Char	1
turn off room lights	Char	2
turn on the fan	Char	3
turn off the fan	Char	4
turn on the main lights	Char	5
turn off the main lights	Char	6
turn on the T V	Char	7
turn off the T V	Char	8

Once you are happy with the control phrases, press **Start** and test it out. Do the relays obey their master? If so, well we are done with all the tests.

Now, we will complete the circuit.

Complete home automation

We are finally here. This is the last time we will be calling our electrician friend to help us. Depending on what appliances you are actually using as compared to what is being said in this section, the circuit arrangement may differ. However, the logic is the same.

Unplug the UNO. You will have to more or less create the circuit that is on the next page. Take a look at it. Do not let the number of wires confuse you. All the wires are insulated. The only wires that are connected to each other are the ones at the main line and the ground (neutral) line. The relay (connected to the Arduino) needs to be placed in a location close to the router, but it is also convenient enough to connect all the wires.

It is recommended that you keep your Arduino + CC3000 + relay setup close to at least one switchboard (perhaps the one in your room). The following image is just to show you where and how the connections go, but it is not an exact representation of how it will be. Physically placing the Arduino board close to a switchboard will make the wire-connections to that switchboard and the relays short and neat. All non ground wires from an appliance are connected to one side of the switch on the switchboard. The other wire connected to a simple switch will be the main (power).

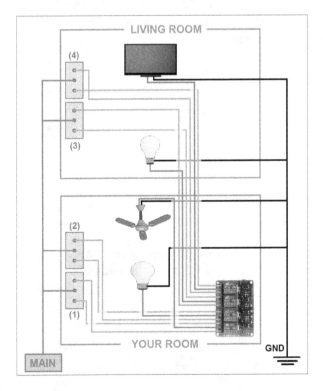

Each relay has to be connected to a two-way switch, which would mean that you would have to replace a simple switch with a two-way switch for this setup to work. As before, be very careful while handling high voltages. Be patient and take your time. If you are controlling a single room, you will not have wires going around, but if you want to control appliances in different rooms, this is the most feasible option. Alternatively, if your budget allows, you can get another UNO and a CC3000 shield and connect them near another switchboard.

Power the Arduino through the 9V 2A power plug and let it connect to the network. Begin speech recognition and start the app on your smart device. What are you waiting for? Press, tap, and speak away! Feeling Tony Stark-ish yet? If not, you should. This is a milestone that you have reached! It was a wonderful moment when I first got this to work some time before working on this book. It's truly spectacular and you may feel like you can do anything and build anything now.

To take things further, you can buy yourself an 8-channel relay and an Arduino MEGA board, which would allow you to control even more appliances. You can also get yourself a wireless microphone and place it in a convenient spot; connect it to your computer so that you neither have to be close to your computer nor have to yell, for the speech recognition software to work.

Summary

Although no summary can do justice to the amount of knowledge you have obtained from this chapter, let us try our best. After learning, in the previous chapter, about how relays function and how the CC3000 works, we programmed them to work in unison. We were able to control the relay and in turn a home appliance by using a terminal on our computer and smart phone. In addition to that, we used some apps to directly control the relay through aesthetically pleasing GUIs. Then, we learned about speech recognition and understood how accurate (or inaccurate) it can be. We then used it to control the very same relay. Finally, we put together everything we have learned, in order to control four home appliances (with the help of our electrician friend), using the terminal, smart phone, and speech recognition. Finally, we built a complete home automation system.

In the next chapter, we will be building and programming a robot dog. This is one of my favorite projects because you can program any sort of personality as you would like. Since we have created a smart home, we will need a guard too, right? The next project will involve using a lot of new parts such as servos to aid motion, and will be quite different from this project. I hope you enjoyed this project and that you are confident enough to modify this project to suit your needs.

8
Robot Dog – Part 1

We've finally made it to the last project. More than the destination, I hope the journey so far has been enjoyable and educational. The previous chapter was on relays, relays, and more relays. Home automation is a very much an upcoming field, and I am glad that we had the opportunity to build a project that helped us to learn about it. We learned to communicate through Wi-Fi via the CC3000 Arduino shield. Using this, we were able to control the Arduino using our smart device, and ultimately we controlled all electrical home appliances. We also used BitVoicer, running on the computer, to enable speech control.

Now, coming back to this chapter, we will be building a robot dog. Cool! That's right. This is one of my personal favorites. We will be using everything that we have learned so far in order to create a small quadruped (four-legged) robot with a lot of capabilities, which you will learn about soon. You are warned that this project is very hard. *Why?* This is because you will be working with many servos (think of them as little modified motors) — the number of wires running around will make you dizzy, and the battery that is very powerful, could cause significant damage if not cared for properly — and programming the robot is a pain. *Why should I even start this project then?* Don't let the complexity demotivate you. I just want you to be mentally prepared before beginning this project. Besides, we will be building this together, and you will be guided every step of the way.

This project, **Robot Dog**, is split into three chapters to balance the length of each chapter. The project is further split into the following sections:

- Introducing Arduino MEGA
- Understanding servos
- Understanding power requirements
- Building the chassis

I have given you my word that you will always be learning something new.
In this last leg of the book, we will learn about using servos. Servos are, in simple words, motors that can only rotate from 0 to 180 degrees. They are the most commonly-used mechanical components for physical motion with Arduinos. Since they are digitally-controlled versions of motors, they are easy to program and can also be used to do different tasks. We will also learn to build the robot from scratch, by using our own creativity and not using an assembly-ready kit. For the first time in this book, we will not be using an Arduino UNO, but instead migrate to its big brother, the Arduino MEGA. Don't fret. It is exactly the same as the UNO, except it has many more input/output pins that allow us to connect more components. This project is very detailed, so be prepared.

Prerequisites

To minimize the cost of this project, while also trying to teach you that sometimes household items can be used in creative ways as materials for a project, we will be using ice cream sticks to create the chassis (ice-cream sticks should be available in your local stationery store; if you, however, choose to acquire the ice cream sticks by visiting your local ice cream parlor every day and get sick of doing so, I am not responsible; nevertheless, this is the way to go).

The rest of the components needed for this project are as follows:

- 1x Arduino MEGA 2560
- 1x USB cable A to B (also known as the printer cable)
- 12 x 9g Micro servos (2+ extra servos recommended)
- 1x Breadboard
- 40x Male-to-male connecting wires
- 20x Male-to-female connecting wires
- 20x Ice cream/popsicle sticks
- 1x Wood glue
- 1x Ruler/measuring tape
- 1x Regular-type screwdriver
- 1x Multimeter
- 1x 7.4V (2 Cell) 2200 mAh (or above) LiPo battery
- 1x 3A UBEC (Universal Battery Elimination Circuit)
- 1x XT60 male
- 1x 7.4V 500 mAh LiPo battery
- 1x Female JST connector

- 1x LiPo battery charger
- 1x Pack of rubber bands
- 1x Pack/roll of double-sided tape
- 3x Insulating tape (different colors)
- 1x PC with a microphone
- 1x Proto board
- 40x Male headers
- 1x Soldering kit
- 1x Thumb pin or needle
- 1x Pack of paper clips
- 1x Pliers
- 1x Wire cutter
- 1x Cutting blade

You will also need the following software:

- Putty (terminal)
- BitVoicer (speech recognition)

Since the shape and size of ice cream sticks will be different in different parts of the world, there would be the need of strengthening the sticks, by sticking two or more of them together. Hence, it is suggested that you buy more sticks.

To give you an edge for planning purposes, here is the ice cream stick used in this project, along with its dimensions:

Introducing Arduino MEGA 2560

In this section, we will briefly go through the Arduino MEGA 2560 board. We will understand how it is different from the Arduino UNO, and also see why we chose this board over the Arduino UNO.

The microcontroller

As mentioned before, Arduino MEGA 2560 is like a big brother of Arduino UNO. It is almost twice as long and way more powerful; Arduino MEGA 2560 is the successor of Arduino MEGA. Unlike Arduino UNO, MEGA 2560 has 54 digital input and output pins, as shown in the following image:

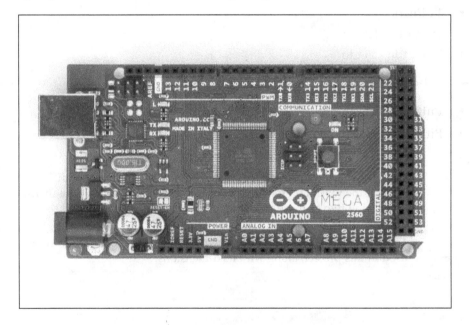

With the bottom looking like this:

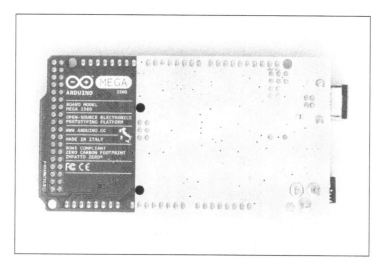

Before we get started with using the board, it is a good idea to use one side of the double-sided tape to cover the base of the board. This is to ensure that, while in use, the Arduino does not get short-circuited if the base makes contact with a conductive material. Stick a double sided tape to the bottom of the Arduino like this:

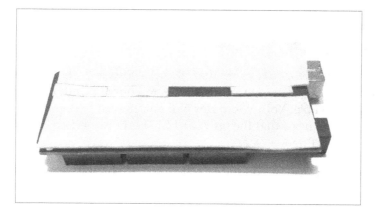

On the right side of the MEGA (short for Arduino MEGA 2560), you will notice some missing labels. The following image should help get this clarified:

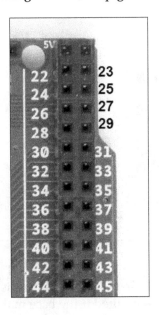

In the pin-pair shown in the preceding image, pins 22 and 23 are 5V pins, just like the one in the Arduino UNO.

Testing MEGA

We will quickly test MEGA to ensure that it works as expected, and to also give you a feel of the board. As always, plug MEGA to your computer by using the corresponding USB cable. Your computer will automatically search for the drivers and install them. If you see that the on-board LEDs are turned on, you know the board's connection is fine.

Launch Arduino IDE and, as you did in the very first chapter, open the **Blink** example. It can be found in **File → Examples → 01. Basics → Blink**. Before uploading the program, you need to select Arduino MEGA as the board to be uploaded to.

To do this, go to **Tools** → **Board** → **Arduino MEGA 2560**, as shown in the following image:

Don't forget to choose the right serial port too. Only after that, you can upload the program. You should notice the on-board LED blinking. Once you have verified that Arduino MEGA works as expected, you are ready to move on.

We will go to the next part where we will learn about servos.

Understanding servos

Just as we had done for the relays in the previous chapter, this section will tell you what a servo is and how it is used. We will also test the servo by using a **Blink** like example.

Servo 101

A servo (or servo motor) can be thought of as a motor, but with the ability to be controlled by its angular position. The servos we are using are 180 degrees micro servos, as shown in the following image:

Servos come in different sizes and capacities. We could use the standard-sized servos that would give us more torque, but it would require more power. For this purpose, a micro servo will do the job.

In the preceding image, you may have noticed that there were three wires. The wires' colors correspond to the following:

- Black/brown: GND
- Red: 5V
- Yellow/orange: Signal

A signal is basically how we control how much a servo turns (between 0 and 180 degrees). Let us now test one.

Testing a servo

Pick up a servo. It should come with different types of servo arms (the typically white-colored pieces that can be attached to a servo and then screwed together). For now, choose any arm and fix it onto the servo. For this part, you need not screw it. However, if you think that the attachment is too loose, you can gently screw them together. The servo will look like this:

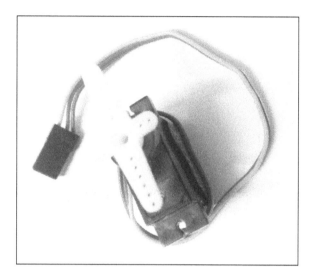

Try to physically, but gently, turn the arms. You will notice that you can freely turn the arm until you reach a certain point. This is a barrier that does not allow the servo to turn more than the preset 180 degrees. If you turn it in the other direction, you will notice another barrier. You might also notice that the two barriers are not located exactly 180 degrees apart. Do not worry about this. If you couldn't budge the arm when you first tried it on, you know that there is something wrong with your servo. Perhaps the gears inside are broken. You will either have to replace the gears, or simply use a new one. It is a good idea to test all the servos that you have purchased before proceeding.

In the following image, you will notice that, at the very centre, there is a white protrusion from the base of the top big gear. This protrusion is marked here:

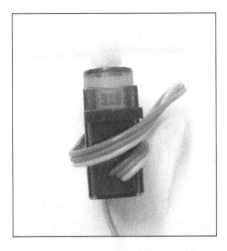

Try moving the arms of your servo to achieve this position. This is the central position of the servo (90 degrees).

Programming a servo

Programming a servo is almost as easy as programming an LED. Yes, these are the exact same words that I had used when introducing you to programming a relay in the previous chapter. Disconnect the MEGA USB cable. Go ahead and create the following circuit:

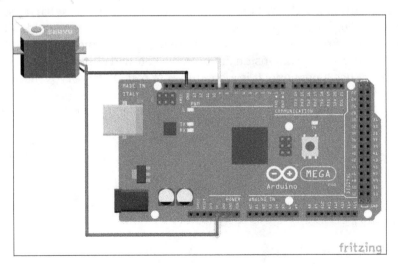

You will have to use the male-to-male connecting wires to connect the servo to MEGA.

The connections from Arduino to the servo are as follows:

- GND → GND (black/brown)
- 5V → VIN (power — red)
- D09 → Signal (yellow/orange)

Plug MEGA back in. You may hear some sounds from the servo. It may even change its position. This happens, do not worry about it. Now, open the servo sweep program, which can be found at **File → Examples → Servo → Sweep**. Then, upload the program.

You will see that the servo rotates from 0 to 180 degrees and comes back to 0. It keeps repeating this action. In this program, look at the following segment inside the void loop:

```
for(pos = 0; pos < 180; pos += 1)   // goes from 0 degrees to 180
degrees
  {                                  // in steps of 1 degree
    myservo.write(pos);              // tell servo to go to position
in variable 'pos'
    delay(15);                       // waits 15ms for the servo to
reach the position
  }
```

Notice the line that mentions `delay(15)`. This is a very important line of code. This `delay` controls the speed at which the servo turns. For now, it is a good idea to keep the minimum delay at `15`. You can try adding another servo, and change the program accordingly to get the similar result. If you need some help, read the following section.

Using multiple servos

Let us now try to control multiple servos. This is our ultimate task to make the robot walk, isn't it? Since the Arduino MEGA has only three 5V pins, we will make use of a breadboard to make the connections. Create the following circuit:

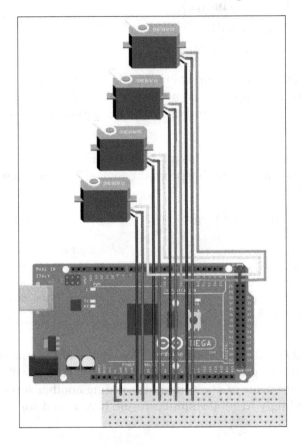

The diagram depicts only four servos. Use the same pattern and connect all the 12 servos. As I said before, the number of wires running around is dizzying. Be careful while using the male-to-male connecting wires. Make sure that the ground, voltage, and signal wires go only to their respective destinations.

The connections are basically as follows (Arduino → Servos):

- GND → GND
- 5V → VIN
- D22 → Signal Servo #01
- D23 → Signal Servo #02

    ..

    ..

- D33 → Signal Servo #12

Now, open the `multi_sweep.ino` file that should have come with this chapter. Plug in your MEGA and upload the code. What do you see? If the Arduino MEGA didn't *die*, then you are one of the lucky ones. In most cases, you will notice that the Arduino MEGA simply turns off, turns itself back on, tries to run the program, fails to do so, and turns off again.

Do you know why this happens? The answer is, because the MEGA simply cannot handle the load of all those 12 servos running at the same time. At most, you can get away with running four servos at a time. The total current required by all the servos when they are in action is simply not being supplied by the computer's USB port. *You didn't just kill my MEGA, did you?* No, don't worry. The Arduino MEGAs are made to withstand much more than the insufficient current. How are we going to deal with this then? How can I control 12 servos, when I can't even power them?

This is exactly what we are going to discuss in the following section.

Understanding power requirements

The basic necessity for any electro-mechanical device is a power source. Selecting the right power source for a particular device is an important trick of trade that every **Arduino tinkerer** needs to know. This is exactly what will be taught in this section.

Limitations of Arduino MEGA 2560

If you look at the specifications of Arduino MEGA 2560 on the Arduino website, you will see the following:

Microcontroller	ATmega2560
Operating Voltage	5V
Input Voltage (recommended)	7-12V
Input Voltage (limits)	6-20V
Digital I/O Pins	54 (of which 15 provide PWM output)
Analog Input Pins	16
DC Current per I/O Pin	40 mA
DC Current for 3.3V Pin	50 mA
Flash Memory	256 KB of which 8 KB used by bootloader
SRAM	8 KB
EEPROM	4 KB
Clock Speed	16 MHz

The important thing to note is the operating voltage that says 5V. This means that no matter how large of an input voltage you put in, the MEGA will always convert it to 5V. In reality, it will be slightly less than 5V due to miscellaneous resistances. Another thing to note is the DC current per I/O pin that says 40 mA. An average micro servo has a voltage rating of 4.8 - 6V, and under heavy load, its current consumption can reach up to 1A that is 1000mA. So, it is surprising that MEGA could even power one servo sufficiently.

Choosing the right power source

Since we are making a robot dog and not something that is stationary, we are going to rule out using a wall power supply such as the one we used for the home automation project. Also, since a computer cannot provide sufficient power to the Arduino, this is also crossed out of the list. This leaves us with batteries. However, there are so many batteries to choose from. Obviously, the component list for this project, at the beginning of this chapter, has already given away what we are ultimately choosing; let us understand what our choices were and how we decided what to go with.

The reason why we are going through the process of finding the right power source is because this is very important, not only for this project, but also for your future endeavors.

Okay, so we are going to use batteries. Yes. *What kind?* Since we are going to build a robot and we want to increase reusability, we will cross out nonrechargeable batteries. This leaves us with two popular choices:

- Rechargeable AA battery packs
- LiPo batteries (Lithium polymer)

As mentioned before, a micro servo requires 4.8 - 6V to work. Each servo also has a peak current draw of 1A, which makes it 12A in total. In reality, not all servos will be moving at the same time, so the 12A is just a maximum figure we are using for calculation purposes.

Each rechargeable AA battery has 1.2V. So, if you connect five of them in series, you would get a total of 6V that would suffice. Now, you also need to make sure that the current provided by the batteries is high enough for the servos to work. Rechargeable AA batteries rated ~2000+ mAh batteries are readily available in the market and can also be used. The reason why we are using LiPo over AA batteries is because the former lasts longer and their performance to weight ratio is much greater. However, if your future project doesn't have to take weight into account, you are free to use rechargeable AA batteries.

Now, let's talk about LiPo batteries. As mentioned in the prerequisites, we will be using a 7.4V (2 Cell) 2200 mAh LiPo battery. If we look at the specifications for this battery, we will see that the constant discharge rate is 20C. '20C' means that the battery can give off 20 times the current of its rated capacity. This means that the battery will discharge 20 * 2200 / 1000 = 44 Amps at a constant rate. This is safely above the 12A required by the servos. The more the mAh, the longer the battery will be able to power the Arduino. A LiPo battery with an ever high mAh could be chosen, but that would increase the weight of the battery. Since the battery is going to be carried by the robot, increasing the battery's weight increases the load on the robot. Also, there is only so much load those micro servos can endure. So, we have to create a balance between performance and weight. Hence, the 7.4V (2 Cell) 2200 mAh LiPo battery was chosen.

Using the right power source(s)

If you had noticed that there were two batteries mentioned in the prerequisites, you were correct. The first and more powerful one that we just talked about is going to be used only for the servos. The smaller 500mAh battery will be used only to power the Arduino separately. The reason why we are not hooking up the 2200mAh battery to both the servos and the MEGA is because of the noise that is generated by the servos, which might cause unwanted disturbances in the MEGA.

However, didn't you say the servos are rated at 4.8-6V? You now want to hook up a 7.4V directly to it? Hold your horses. This is what we are ultimately going to do, but before that I must introduce you to our little friend, the 'UBEC'. What this device basically does is, it drops the voltage (no matter how high it is) to 5 or 6V, depending on what is wanted without compromising the current. Since we are going to stick with 6V, move the jumper to choose 6V as the output. The UBEC should now look like this:

Solder the voltage and ground (red and black respectively) wires to the XT60 male connector. Make sure that the ground is soldered to the terminal with the chamfered side. Wrap some insulation tape around it to make sure no unwanted connections occur. On the other side of the UBEC, the triple wires that look like that of a servo will basically act the same. Meaning, the black/brown wire represents ground and the red wire represents the constant 6V. The yellow/orange wire is the signal that is usually used in RC planes in receivers, but we will not make use of this wire.

Now, we are going to modify the previous circuit to ensure that all the servos work together without choking the MEGA. To do this, create the following circuit:

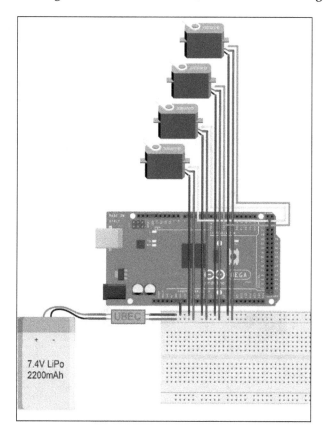

The main thing to notice is that only the servos are being powered by the LiPo + UBEC. The MEGA is still being powered separately. Ensure that the right polarities are connected to and from the battery and the UBEC. One key thing is that the ground of the breadboard should be grounded to the Arduino MEGA. This is to establish one common reference ground. Be very careful while handling the high discharge battery. Now, try running the `multi_sweep` program again. See the difference? The servos don't stop; they keep rotating. This means that their power requirements have been satisfied. If a particular servo is not rotating, make sure that its wiring is done properly.

Before going to the chassis construction part, connect the smaller 500mAh battery to the 2.1mm DC plug by using the female JST connector. You'll have something like the following:

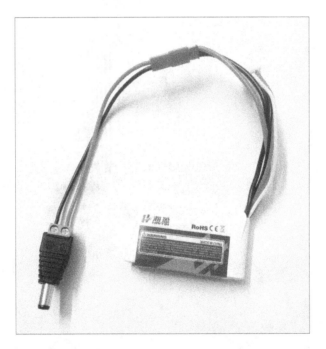

Now, unplug the MEGA from the computer and plug the 2.1mm plug into the MEGA's 2.1mm jack. It works! So, you have successfully used a portable power source to power the servos and the MEGA independently.

Building the chassis

A robot always comprises of three fundamental fields. They are mechanical, electrical, and programming. Working out the power requirements falls into the electrical aspect. Writing code to control the motion of the servos falls under the programming category. Building and ensuring the chassis can support the weight of all the components that the bot is carrying is a mechanical challenge that will be addressed in this section.

Using prior art

Before rushing into the building phase of the body of the robot, let us take a moment to understand what we are even trying to build in the first place. We are making a robot dog. So, let us look at the nature and see how a dog stands and moves.

It is very hard for us to replicate the dog at its entirety. This is mostly because an actual dog uses strong muscles in its legs to walk (jump), and we are using servos. Now, let us look at the following bone anatomy of a dog to get a deeper look at the skeletal structure that allows the dog to walk the way it does:

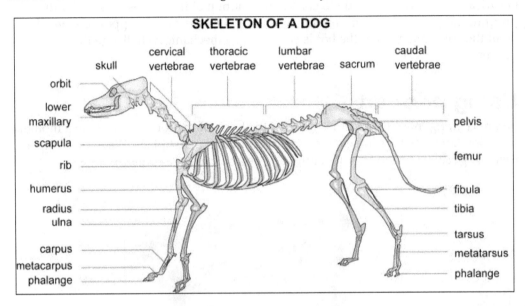

Why are we doing this? Didn't I say that I will teach you as much as possible? This part of the project is called *getting inspiration from prior art*. This technique is used almost everywhere while designing a new product or simply creating a new design. It is used in research papers and even patents. Sometimes it helps to use what has already been done before, either natural or artificial, to aid in creating a new design.

Let us focus on the legs. Notice how the joints are connected. To make things easier, we will simplify this anatomy to fit our needs. We will modify the legs to just have a joint at the hip and the knee. This ensures that we won't have to use too many servos. Also, we are going to make the body (not including the legs) flat.

Since we are using straight, flat ice cream sticks as the bones, we have to ensure that they are proportionate. Here is a basic side view of what we will be making:

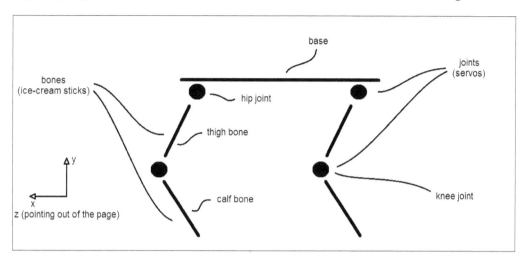

This stick figure shows the side view of our robot. The bones will be made of popsicle sticks (ice-cream sticks), and the joints are simply servos. The base will also be made of the same sticks, but it will be rectangular in shape, with its corners attached to servos; the central part will be made to support the other components such as the Arduino, batteries, and so on.

The hip joint (closer to the base) will comprise of two servos attached to each other so that the leg can be rotated in two axes (axis): the y-axis and the z-axis. The knee joint (conjunction of the thigh bone and the calf bone) is going to have a single servo that allows the calf bone to be rotated only in the z-axis. Make a note of all the names of the joints and the bones in the preceding diagram. We will be using the same names during the construction and programming process, to label different joints/bones.

Now that we have a basic understanding of what we are going to build, we can proceed with its actual construction that will be covered in the next chapter. You are free to change some details of the structure while we build it. However, you should be wary that this change may result in intricate changes you have to do for the code as well.

Summary

This chapter is very different from the previous chapters that we have gone through. This is because, in this chapter, we looked at every detail to ensure that the learning process aids in your future projects. We started by introducing ourselves to the Arduino MEGA 2560 and even ran the **Blink** example, which we had executed a long while ago in the very first chapter with Arduino UNO. After that we learnt about servos. What they are and how they are programmed. While programming multiple servos, we faced issues that were concerned with the power requirements for the servos. In the subsequent section, we learned how to select a suitable battery to nullify the issue. We even tested this out and were able to successfully control all the 12 servos through the Arduino MEGA. Then, we imitated the process of building the chassis by looking at the prior art of dogs, and created a simple sketch that helped us visualize what we are ultimately going to build.

In the next chapter, we will actually build the entire chassis by using ice cream sticks, and fit the joints with servos. We will also complete the circuit that will allow us to control all the 12 servos.

Robot Dog – Part 2

9

The previous chapter introduced you to the Arduino MEGA: servos and battery requirements. It ended with an understanding of what we are going to build by the means of using a dog as prior art.

In this part (2 out of 3), in the series of the robot dog chapters, we will focus on building the chassis of the robot dog, and then completing the circuit. What we are about to do in this chapter is time consuming, so it is okay to take breaks in between to give your fingers and mind some rest. In summary, we are going to cover the following topics:

* Building the chassis
* Completing the circuit

That being said, let us begin.

Building the chassis

We have already finished the *prior art* part of this section. That subsection has inspired us to design our robot in a similar fashion of a dog. We will use this knowledge to build the body of the robot. This section may bring about nostalgic memories of art and craft projects that you did when you were little.

Sticks and servos

Start by clearing out some space on your workspace. Then, pick out two working servos and put on their arms.

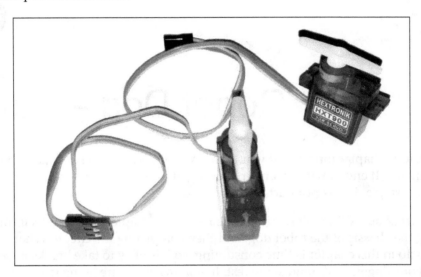

Now, cut out a piece of double-sided tape such that it covers the base of the servo as shown:

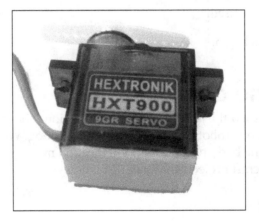

Now, stick this servo onto the other servo at a 90 degree angle like this:

Use two rubber bands and stretch them around the servos, as shown in the following image, to strengthen their bond:

This is going to be our hip joint that was mentioned earlier.

Now, pick the straightest-looking ice cream stick, mark a line that is 7 cm from one end, and cut it:

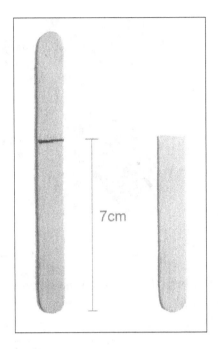

Now, laying the two-sided arm of the servo at the edge of the cut stick, poke two thumb pins (or any other pin) such that one of them is at the center and another one is through the tiny hole along the arm of the servo, like this:

Now, remove the pins and the servo arm. Use the pins to make holes through the thickness of the stick where they were previously poked. It is suggested not to hammer the pin into the stick because this may result in creating a crack along the length of the stick from the hole, which decreases the strength of the stick. Use a smooth rotational motion to push the pin through the stick.

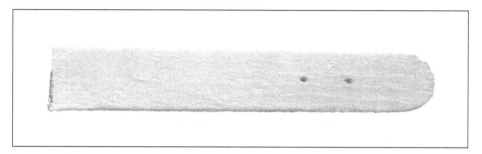

To attach the servo arm to the stick, we will use a screw that goes along the axis of the servo, and also a paper clip. Most paper clips, such as the one shown in the following image are metallic with a plastic coating:

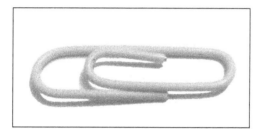

We only want the metallic 'wire' that lies inside the coating. To get it, unbend the paper clip to make a straight line. If it is hard to accomplish this with your fingers, you should use a plier to do this job:

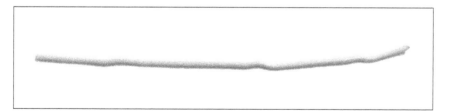

Now, holding a plier in one hand and a wire cutter in another, remove the coating of the paper clip:

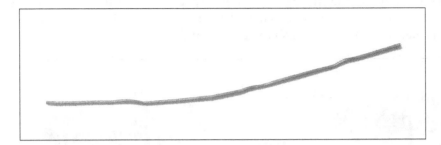

Fold this clip into half and cut it into two pieces at the fold:

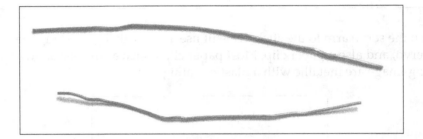

Place the arm of the servo at the position corresponding to the holes that we made earlier. Push one of the halves paper clips through the hole on the arm (not the central one). Fold the clip in such a way that it tightly locks the arms in place. Use pliers to do this; be careful of the sharp edges:

Bend the straightened clip to fasten the servo arm to the stick:

Now, center the lower servo. Attach the servo to its servo arm counterpart, such that the stick is perpendicularly downwards:

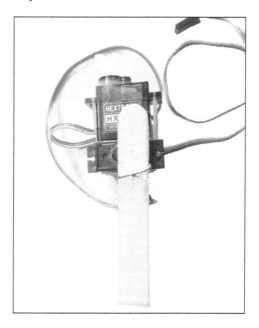

Once you are satisfied with the positioning of the stick, tightly screw the stick to the servo using a screw that came with the servo. These are the approximate angles you could achieve with this arrangement:

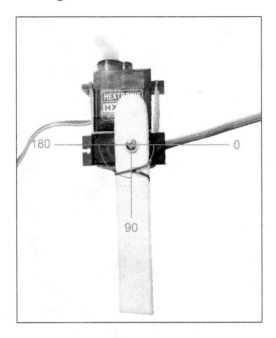

Now, take another servo and apply a double-sided tape to cover its base. Stick this servo to the stick above, in such a way that the edge of the servo superimposes on the edge of the stick:

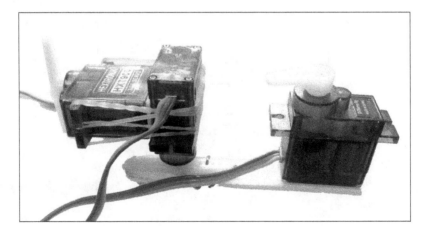

Now, tightly bind this servo with rubber bands as well:

This servo serves as the knee joint. Now, take another ice cream stick, measure 6.5 cm from one end, and cut it at that mark. As we did before, take a servo arm and place it at the edge of the stick. This time, place it at the blunt edge. Use two pins to mark the locations of the necessary holes:

Make the holes using the pins, and attach the servo arm to it. Hold it in place by using the other half of the unsheathed paperclip. Center the knee servo and attach the stick in this manner:

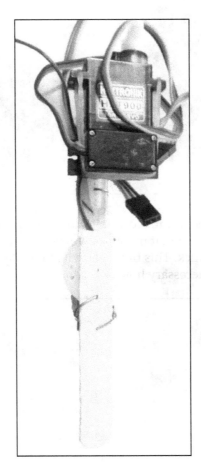

When you are okay with the angular movement of the stick, screw it tightly. You should be able to get the following angle with this joint:

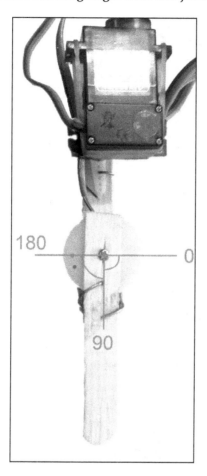

You have now created one of the four legs. There are three more to go. You can create another leg in exactly the same way as you did just now.

Next, you need to create the other pair of legs. The only difference is about the way the servos at the hip are joined together:

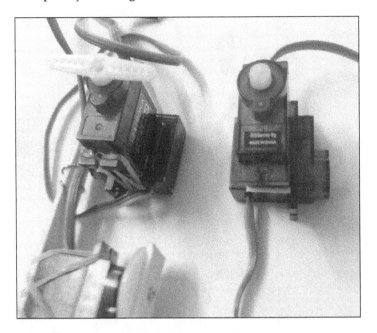

They are mirror images of each other. Now, create two legs with that joint. When you are ready with all the four legs, we will be able to move on:

Now, we need to create the base. Take four ice cream sticks with the same length and place them in the following formation:

Basically, each stick is placed such that its 'curvy end' is just outside the superimposition region. Go ahead and stick them in this position with wood glue. What you can do now to make the base a bit stronger is lay more ice cream sticks in the following manner and stick them:

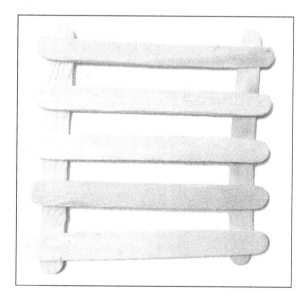

Now, the base is ready. All that is left to do is attach each leg to the corners of the base. To do this, mark points 2 cm away from each end. As we did before, place a servo arm and hold it in place using two pins, create holes in these points with the pins, and send an unsheathed paper clip through the noncentral axis hole and bind the arm in place. Do this at all the four corners and you'll get the following result:

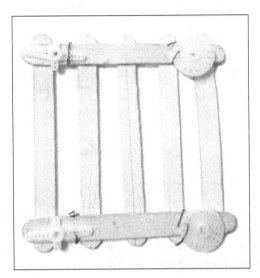

In your case, the arms will look identical. Do not worry about how they look. In this project, I chose different arm shapes to differentiate between the front and back servos. You will soon realize that the number of wires running around will get real confusing, real quick. Now, all we have to do is attach the legs. For our purposes, we are going to make the calf bone (stick) outward. This should give the robot enough balance to stand.

Now, center the servo and attach the leg as shown in the following:

Do this for all the other legs and you'll get the following result:

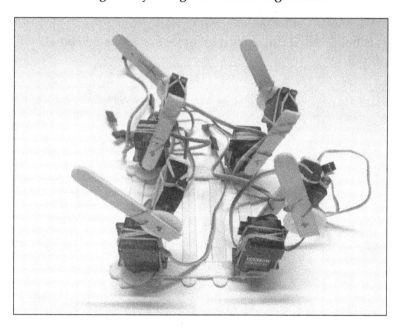

The following shows the standing position:

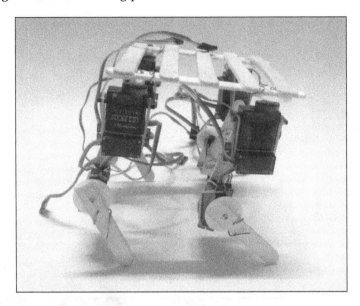

If this chassis is built well, it can carry a lot of load, such as this bowl:

And there we have it: the completed body of the robot.

Completing the circuit

Now that we completed building the chassis (most of it at least), we can move on to the electrical part of the project. This involves connecting all the servos to the Arduino MEGA; this is exactly what this part entails. "However, didn't we already do this in the previous section?" Yes, but to be more effective, we are going to firstly label the servos and create a circuit that drastically decreases the number of wires that we need.

Labeling the servos

Just before connecting all the servos together to the MEGA, let us make the life of our future selves easier. Let us label each servo so that later we can easily identify what servo wires connect to what servo. To gain an understanding of this, take a look at the following image:

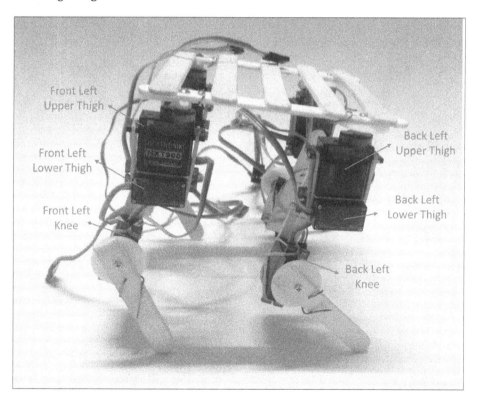

A symmetric labeling will follow on the right-hand side. We will use the following chart to label each servo. The number of (bands refer to the thin strips of the insulation tape that is used as a type of identification):

Servo location	Color	Number of bands
Upper thigh	Black	1
Lower thigh	Black	2
Knee	Black	3
Front	Red	1
Back	Red	2
Left	Yellow	1
Right	Yellow	2

Doing this for the entire servo would look like this:

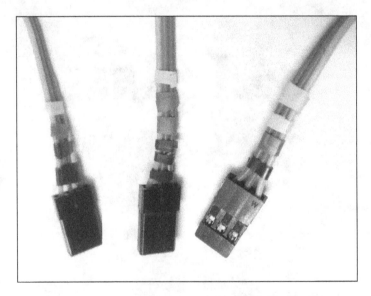

It might have been a bit frustrating to do this, but once you are done, it will be very easy to identify and replace a faulty servo later. It also aids programming and such.

Building a tiny circuit

This is something new. As mentioned at the very beginning of the chapter, we are going to create a circuit to remove the need for the breadboard and drastically reduce the number of wires needed for the servo connections. For this, we are going to use a proto board.

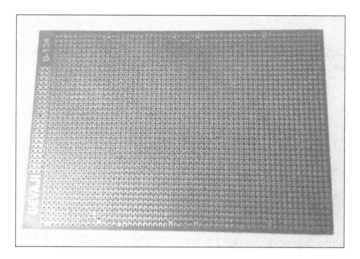

Of course, we do not want such a large proto-board, so we will have to cut it. We will use a 4 x 15 size board, where 4 and 15 represent the number of column holes and row holes, respectively. To cut it, use a straight edge (as shown the following image) and use a blade to cut through the board. It may not cut this way, so turn the board to the other side and repeat it. If it still doesn't cut, use pliers to carefully separate the proto-board into two pieces:

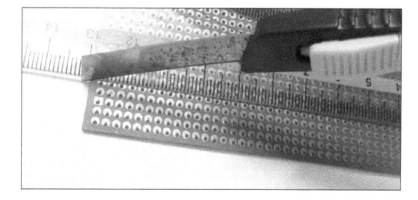

The cut piece will look like this:

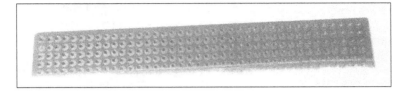

Repeat the step again to decrease the number of rows to 15, as shown in the following image:

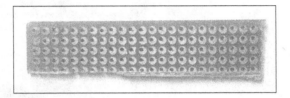

Now, pick up male header strips and snap them into 1x14, 1x13 and 2x12 pin lengths. If you don't have enough to create a set of these numbers, it's alright; you could always merge 7 and 5 pins to make 12:

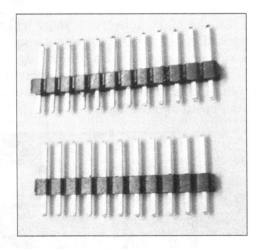

Place them into the 4 x 15 cut proto-board as follows (ensure that the shorter part of the header is on the side, with the little copper circles):

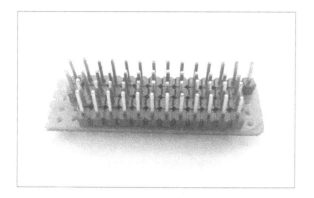

Now, solder the underside (copper circles) in the following manner:

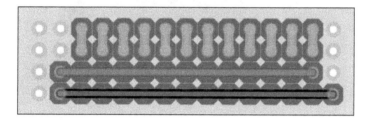

The bottom will look like this:

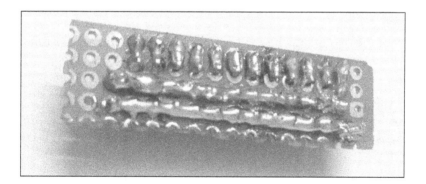

I hope you understand what we are doing with this tiny circuit. We simply created a more efficient way to connect all the servos through the length of the tiny circuit and connected the battery directly to this board. To test the soldering, you can either use a multimeter set to measure resistances, or use the MEGA. The former is self-explanatory. For the latter, remove all the connections that go to the MEGA and create the following circuit:

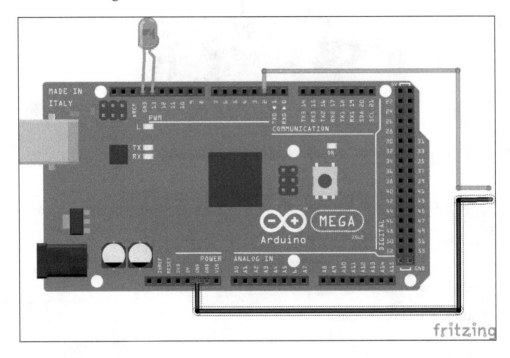

Make sure that the two unattached wires that come out of the MEGA have female terminals. Open Arduino and launch the **Button** example that can be found via **File → Examples → 02.Digital → Button**. Then, upload the program.

The LED on pin 13 will stay ON as long as no connection is detected (the circuit is open). Now, use these two wires to go through the tiny circuit to ensure that the soldering is done properly and the connections are as required. Who needs a multimeter, right?

Now that we have built the circuit, we just have to complete the connections.

Putting it all together

If you haven't yet appreciated the advantage of the tiny circuit, you will now do so. Connect the edge with the 12 pins of the tiny circuit, with 12 male-to-female connecting wires. Just before going to the connections, let's take a look at a few abbreviations that will help us:

- UH: Upper Hip
- LH: Lower Hip
- K: Knee
- F: Front
- B: Back
- L: Left
- R: Right

Now, here is how the rest of the connections will go about with the configuration being components → tiny circuit → Arduino:

Components	Tiny circuit	Arduino
F R UH	#1	D22
F R LH	#2	D23
F R K	#3	D24
F L UH	#4	D25
F L LH	#5	D26
F L K	#6	D27
B R UH	#7	D28
B R LH	#8	D29
B R K	#9	D30
B L UH	#10	D31
B L LH	#11	D32
B L K	#12	D33
N/A	#13(GND)	GND
UBEC (5V & GND)	#0 (5V & GND)	N/A

Doing so will result in the following setup:

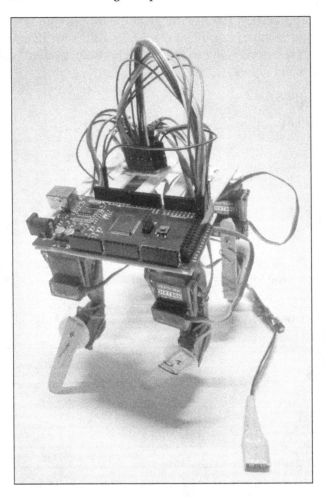

Summary

This chapter was quite something, wasn't it? We planned and built our own chassis for the robot dog using household materials. We painstakingly labeled each servo that our future selves will be grateful for. Then, we created our own tiny circuit, which will help us significantly, by decreasing the necessity of the breadboard and so many extra wires. Finally, we completed the circuit, linking the components, the tiny circuit, and the Arduino MEGA 2560.

In the next chapter, we will program the robot dog to do a series of actions, including standing, sitting, walking, and so on. We will also make some additions such as adding speech control and also adding a personality to the dog.

10
Robot Dog – Part 3

Chapter 10! "Are we there yet?" Finally, yes! In the previous two chapters, we learned about the Arduino MEGA: how to control a servo and supply adequate power; we built the chassis for our dog using ice cream sticks and created a tiny circuit to get rid of the breadboard and a dozen of wires.

In this chapter, we will finally learn to program a complex machine such as the robot dog that we've built, and how to tackle hurdles that we face. The reason why this is so hard to program is because of all the servos that need to be controlled and coordinated in order to produce the right balance and meaningful motions. This chapter is split into the following sections:

- Programming the robot
- Developing a personality
- Implementing speech control

So let us begin part 3 of the robot dog chapters.

Programming the robot

Programming the robot is the hardest task in this book. It is not only about simply writing code and hoping for the best. Technically, you can use that technique and still get away with it, but it would consume a huge amount of time and energy. We will go through the prior art again, see how a dog (and a robot dog) walks, and try to mimic some of that motion. We will also look at the mechanical design of our robot and see how it can be upgraded for better performance.

Weight distribution

Funnily enough, it makes sense to put this in the programming section. This is because weight distribution and the creation of the walking motion go hand in hand. Before programming the bot, we will have to power it. Since we have made our circuit in the previous chapter, we just have to put it in place on the robot, such that the weight of the battery and the Arduino MEGA is evenly distributed.

First, let's take a look at the battery (the larger one):

This is the heaviest object that the robot will be lifting. It will significantly increase the difficulty of programming if it is not centered correctly. Putting it over the base of the robot seems like a logical solution, but the issue with the battery is that it occupies so much valuable space that can be used by the circuits, as they need to be more accessible than the battery. This is why we are going to mount it to the bottom of the robot.

Here is a simple diagram of what the battery's position will look like. Remember, this is the bottom view of the robot:

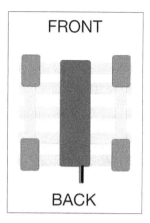

While placing the battery, sort out the servo wires and place them at the corners around the battery. You can mount it using a layer of double-sided tape and then tape to bind it from a place. Ensure that the battery is exactly at the center of the base. A well-bounded one will look like this:

Make a note of how the servo wires are segregated into four parts around the battery. This is to decrease the distance between the servos and the MEGA in order to avoid unnecessary servo-cable extensions. If you think you really need them, especially for the knee servos, feel free to use them.

Arduino MEGA can be mounted directly opposite to it on the top, like this:

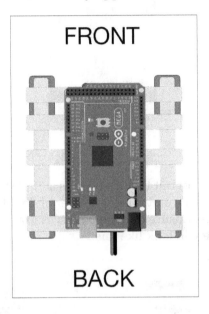

The MEGA can be stuck using a double-sided tape that we had used earlier to cover its base. Again, make sure that the segregated servo wires come out through the corners of the MEGA. For testing purposes, we will leave the rest of the connections as it is for now.

Test one

"Test one?" As mentioned before, you will be taught how to make a robot, including the testing and reiteration phases. We are going to run our first test. It is a simple test that checks the stability of the robot and its ability to stand upright. You can connect the USB to the MEGA. Do not connect the battery yet.

Open the `robot_dog_basic.ino` file in Arduino, upload the program, and carefully connect the batteries to the UBEC. Your robot should spring into life and stand in an awkward stance, or not stand at all.

This is what we are trying to achieve (side view):

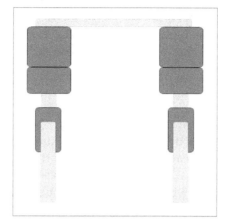

All the bones (sticks) point straight down. If you didn't achieve this, don't be saddened. Let us look at the code and fix it:

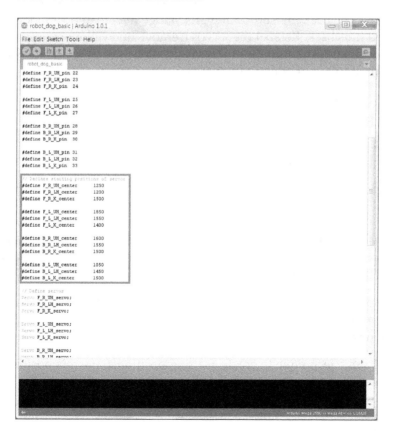

The marked part is very important. The numbers that represent the starting position of the servos used in this chapter and the number you have to use will be different. Try changing one of the numbers, say `F_R_UH_center` (Front Right Upper Hip) right now, and uploading the code. Notice what happens? Tweak the numbers until your robot looks like the required stance shown in the preceding image of the robot.

If you haven't got it standing yet, keep trying until you get it. You can proceed once you are done.

Let's take a look at the robot now. The whole thing looks balanced, but it seems as though something is not perfectly right. It looks like the back legs are taking the majority of the load. This is because we had attached the back hip servos in order to resemble very much like a dog. A dog has more muscles to help balance. However, this slight offset we have here will give us a lot of trouble in the future. *Why didn't you tell me this while we were putting the thing on?* It's simple: nobody knows how to create the perfect project plan. It requires choosing a design and hoping for the best, and reiterating the design on the way.

Unhook the battery and the USB. Unhook all the back servos going to the tiny circuit. Unscrew the back hip servos only. Exchange the left and right legs. Screw the exchanged hips back. Relabel the servos (remove one yellow band from the then right servos and put them onto the current right servos). Connect them accordingly to the tiny circuit following this components → tiny circuit → Arduino:

- B R UH → #7 → D28
- B R LH → #8 → D29
- B R K → #9 → D30
- B L UH → #10 → D31
- B L LH → #11 → D32
- B L K → #12 → D33

Make sure the connections are correct and tight, connect the USB and the battery and check the position the dog now stands in. You may have to change the 'servo center' values to get the desired position.

Save this file or at least the center values, as we will be using it from now on. The dog is much more stable now with the weight equally distributed among the four legs. We can now move to the next step: teaching the robot how to walk.

Try using the same technique to create a sitting position. Save this position and use these numbers when we come across the sitting position again.

The walking gait

Programming the robot's walk is similar to teaching a baby to walk, except that the robot does not have consciousness. A walking gait is basically a sequence of motions that results in the walking action. There are two common gaits: the 'creep/crawl' gait and the 'trot' gait.

The creep gait is what you see a cat doing when it is stalking its prey.

A video (`https://www.youtube.com/watch?v=wQsmsr0oR6c`) would do this more justice, but even from the preceding image or your experience, you can observe how the cat keeps its body constantly horizontal. This is a highly energy-efficient walking gait, but it is very complex to engineer.

The trot gait is when the two diagonal legs move together at a time to create a walking motion, as shown in the following image:

This is not as highly energy-efficient as the creep gait, but is far simpler than the creep gait. Also, this has more static stability, while the creep gait has more dynamic stability. Since you are programming a quadruped robot for the first time, let's take it easy. We will choose the trot gait and slightly modify the motions to make it suitable for our robot.

Test two

We are going to program the robot, using the trot gait. Open the `robot_dog_trot_gait.ino` file in Arduino. Before uploading it, keep the robot in a space where it has no obstacles. A long USB cable benefits here. Change the servo center values to what you had saved in the previous subsection. Plug in the battery and upload the program. Did it move forward or did it fall? Even if you executed a perfect walk, you can glance through the next bit. This may aid you later.

If your robot fell, why do you think this happened? We were sure of the balance. What is going wrong in that case? The ends of the calf bones (sticks) are too thin to cover the area that would further improve stability and weight distribution along the legs. A good way to solve this is using plastic bottle caps. As mentioned earlier, these household materials are easily available.

Again, unplug the USB and the battery. It is not necessary to remove the knee screws to remove the calf stick, but if this is more convenient, go for it. We will need eight bottle caps. If you think four will suffice, then you can choose 4 and run the code again. Mark the center of the two bottles with the thumb pin you used earlier, and push the pin through to make a tiny hole. Tape the bottom part of the calf stick a couple of times to create a washer-like material. Now make a small hole, 1 cm above the bottom of the calf stick.

Now, with one cap on both the sides of the calf stick, drive a screw through the thickness of all the three layers. It is recommended to tape the exterior of the bottle caps to improve traction. It will look something like this:

Do this for all four limbs.

Once done, connect the battery and then the USB cable. You can alternatively use the smaller LiPo battery to power the Arduino at this stage via the 2.1 mm power port on the Arduino MEGA. This would be more convenient because of the length constraint of the USB cable.

It walks, doesn't it? It doesn't fall either, right? It is cool, right? If it does fall, try tweaking these parameters:

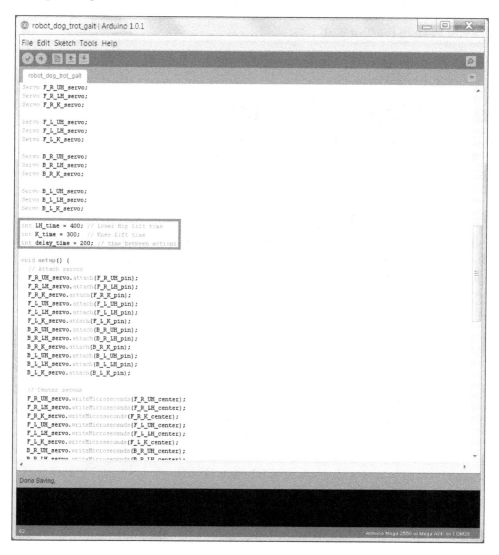

What this program does in sequence and repetition is the following:

- Raise the front-right leg and advance it
- Raise the back-left leg and advance it
- Center the above servos, raise and advance the front-left leg, pushing the robot forward
- Raise the bottom-right leg and advance it
- Center other servos pushing the robot forward

There you have it: the 'trot gait'.

This was the major part of programming the robot.

Developing personality

The personality is what makes the robot more alive. Adding personality to it makes it more fun and likable. This section focuses on some ideas that will help you create your own unique personality for your robot.

Circuit upgrade

The tiny circuit that we made can be split into two; one along each side of the Arduino connected to a common GND and power. Additionally, more header pins can be added to accommodate more servos (the tail and head) in order to organize the wires better.

First, for this, cut out a long circuit board with a row size of about 6 pins.

Following the previous circuit's design, create this circuit:

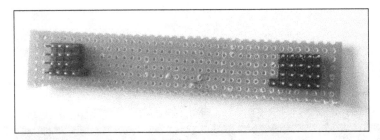

To link the left and right side, small pieces of wires are soldered; one for the GND and another for the power, like this:

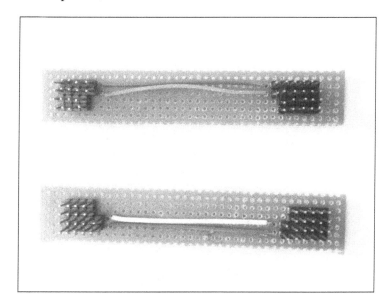

Attach these circuits on either sides of the Arduino MEGA in this fashion:

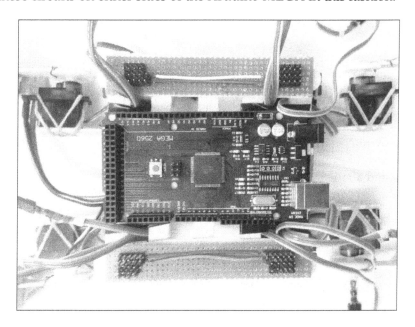

The finished connections will look like this:

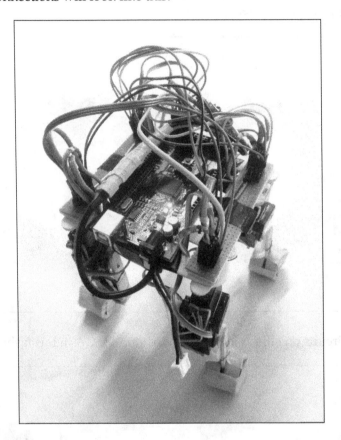

Body upgrade

To be perfectly honest, the robot doesn't resemble a dog at all. Sure, it walks on four feet, but it doesn't resemble a robot dog either. This would require an outer shell with a head and a tail.

Before installing a tail, it's a good idea to create an additional base layer or a case above all the wire commotion. This can either be made of more ice cream sticks, or you can just modify a cardboard box that you happen to have lying around to fit the robot, like this:

This new base can accommodate the tail. The tail can be made using another ice cream stick. It needs to be attached to a servo stuck underneath the base at the back.

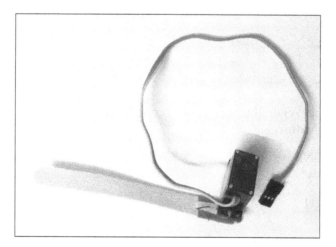

The servo can be programmed to create a wagging effect. You can also attach two servos that would allow you to control the shape of the tail. The tail can be used to express emotions (a wagging tail is a happy tail).

You can additionally attach a make-shift head that would make it seem much livelier. However, in this chapter, we will use something that does give it a face, but not a head.

Sensors

The reason why adding sensors, such as the SRF04 (or SRF04) sensor that you used earlier to measure the distance, is such a great idea is because it enables the robot to get a feel of the environment—only with sensors can it respond to stimuli.

To attach the sensor, make a slit on the top-front part of the base to accommodate the HC SRF04. Place the sensor at the front, as shown in the following image:

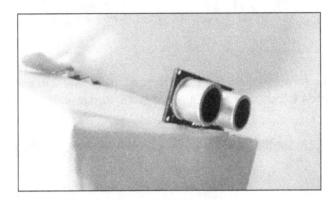

Attach this in place. When the time comes, connect the sensor using the following circuit (SRF04 → Arduino):

- GND → GND
- Trig → D12
- Echo → D11
- VCC → 5V

The switch

Finally, we will connect the secondary battery. To conserve its power, we will make use of a switch like this:

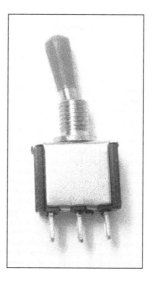

Connect the battery's connector to the switch, like this:

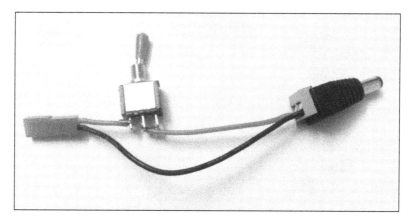

Attach the battery and the setup will look like the following image:

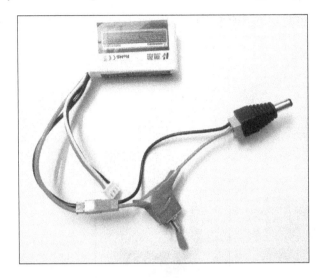

Once you have attached the switch to the battery, attach the battery on the underside of the case. Make sure that you stick it at the exact center of the case:

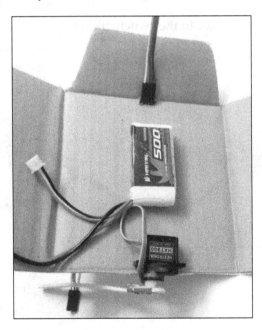

Make a small hole on the side where the wires of the battery are, and attach the switch to it:

Once you are satisfied with the components inside the case, connect all the wires including those for the sensor. Connect the tail to D34. Then, attach the case to the body of the robot.

This is what the final product will look like:

Of course, yours may look entirely different, but that is completely fine.

Coding the personality

Just like the external design of the robot, you can program your robot's unique personality too. Since we have attached the sensor as the head of the robot, we can make use of its capabilities. To give you an example, let us make a 'lazy but loving' robot. Let us briefly discuss the algorithms involved:

- Dog just sits around
- Keeps sitting if not disturbed
- If disturbed (detects a hand in front of its sensors), it stands up and wags its tail
- If the hand slowly moves back, the robot follows the hand
- Ultimately, if the hand comes close enough, the robot will lift its leg to shake its hand with you

Load the `arduino_robot_dog_personality.ino` file, and go ahead and play with it. It is important to note that this program uses both a tail and the ultrasound sensor. If you do not want to use these components, tweak the code accordingly.

Implementing speech control

We have already learned about Bluetooth and speech control in the previous chapters. In this chapter, we are going to merge them.

Connecting the HC-06 module

Just like in the burglar alarm project (Chapters 4 and 5), hook up the Bluetooth to the Arduino MEGA in the following manner:

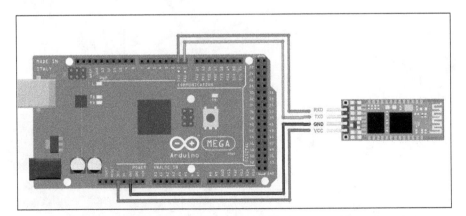

Make sure you are connecting the Bluetooth module to the 3.3V of the MEGA, and not the 5V. Leave the rest of the servo wiring as it is.

Programming the Arduino

Now, open the `robot_dog_speech.ino` file. Just like before, change the servo center parameters to match your robot. Upload the program and connect the computer to the Bluetooth, using the instructions given in *Chapter 5, Burglar Alarm – Part 2*.

Setting up BitVoicer

Open up BitVoicer and use the commands as shown in the following screenshot:

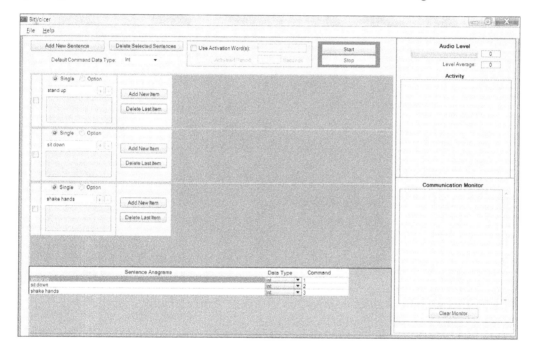

In summary, the commands are as follows (integer type):

- Stand up – 1
- Sit down – 2
- Shake hands – 3

Change the preferences of BitVoicer to match this image:

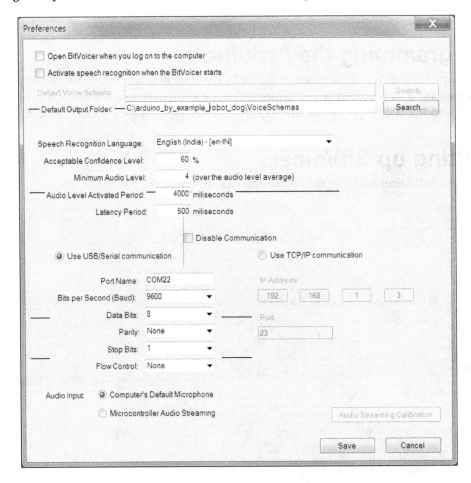

Note that the COM ports for you will be different. You can look it up in the device manager.

Once everything is set, run the BitVoicer schema. Now, you can literally talk to your robot. It won't talk back, but you can make it sit, walk around, shake your hand, or whatever else you want to program. You are free to add additional functionalities via speech.

Unfortunately, you will still need the computer to communicate with the robot. Making the project standalone will require an additional microcontroller, which is beyond the scope of this book. For now, you can, however, use a wireless headset to communicate with the robot wirelessly.

Summary

If you have reached this point, pat yourself on the back. This chapter must have been very exhausting with so many different problems to deal with. However, I hope that at the end, you have successfully managed to get the robot to move the way you wished.

In this chapter, we programmed the robot dog. To further aid our programming trials, we had to change the chassis a bit. We had to do a lot of testing to get things working the way we wanted. While doing so, we learned about the many aspects (and issues) of building a project from scratch. Once we were happy with the basic walking gait, we upgraded the robot. We added a case, tail, and sensor. Using these elements, we created a simple personality with a huge potential for expansion. Finally, we implemented the speech recognition that allowed us to control the robot using our speech. Funnily enough, to complete the revision of the previous chapters, we can also use a relay to act as a switch for the larger battery powering the servos, which can be controlled by the Arduino.

Now, we have reached the end of the book. If you've reached here, you have learned a lot about Arduinos. I hope that the example-based layout of this book, which aids in direct translation of thought from the author to the reader, was helpful. I would like to thank you for having the patience and commitment to go through this book. The only thing left to do now is use the knowledge you have gained from this book and use your creativity to make wonderful creations. Go on, I'm not going to stop you.

Index

Thank you for buying
Arduino by Example

About Packt Publishing

Packt, pronounced 'packed', published its first book, *Mastering phpMyAdmin for Effective MySQL Management*, in April 2004, and subsequently continued to specialize in publishing highly focused books on specific technologies and solutions.

Our books and publications share the experiences of your fellow IT professionals in adapting and customizing today's systems, applications, and frameworks. Our solution-based books give you the knowledge and power to customize the software and technologies you're using to get the job done. Packt books are more specific and less general than the IT books you have seen in the past. Our unique business model allows us to bring you more focused information, giving you more of what you need to know, and less of what you don't.

Packt is a modern yet unique publishing company that focuses on producing quality, cutting-edge books for communities of developers, administrators, and newbies alike. For more information, please visit our website at www.packtpub.com.

About Packt Open Source

In 2010, Packt launched two new brands, Packt Open Source and Packt Enterprise, in order to continue its focus on specialization. This book is part of the Packt Open Source brand, home to books published on software built around open source licenses, and offering information to anybody from advanced developers to budding web designers. The Open Source brand also runs Packt's Open Source Royalty Scheme, by which Packt gives a royalty to each open source project about whose software a book is sold.

Writing for Packt

We welcome all inquiries from people who are interested in authoring. Book proposals should be sent to author@packtpub.com. If your book idea is still at an early stage and you would like to discuss it first before writing a formal book proposal, then please contact us; one of our commissioning editors will get in touch with you.

We're not just looking for published authors; if you have strong technical skills but no writing experience, our experienced editors can help you develop a writing career, or simply get some additional reward for your expertise.

Arduino Development Cookbook

ISBN: 978-1-78398-294-3   Paperback: 246 pages

Over 50 hands-on recipes to quickly build and
understand Arduino projects, from the simplest
to the most extraordinary

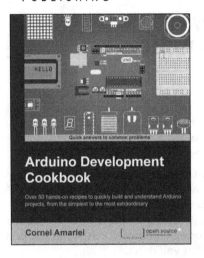

1. Get quick, clear guidance on all the principle
    aspects of integration with the Arduino.

2. Learn the tools and components needed to
    build engaging electronics with the Arduino.

3. Make the most of your board through practical
    tips and tricks.

Arduino Essentials

ISBN: 978-1-78439-856-9   Paperback: 206 pages

Enter the world of Arduino and its peripherals and
start creating interesting projects

1. Meet Arduino and its main components and
    understand how they work to use them in your
    real-world projects.

2. Assemble circuits using the most common
    electronic devices such as LEDs, switches,
    optocouplers, motors, and photocells and
    connect them to Arduino.

3. A Precise step-by-step guide to apply basic
    Arduino programming techniques in the
    C language.

Please check **www.PacktPub.com** for information on our titles

Raspberry Pi Home Automation with Arduino
Second Edition

ISBN: 978-1-78439-920-7   Paperback: 148 pages

Unleash the power of the most popular microboards to build convenient, useful, and fun home automation projects

1. Revolutionize the way you automate your home by combining the power of the Raspberry Pi and Arduino.

2. Build simple yet awesome home automated projects using an Arduino and the Raspberry Pi.

3. Learn how to dynamically adjust your living environment with detailed step-by-step examples.

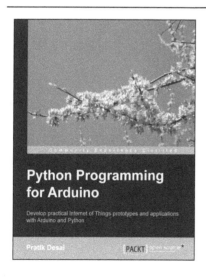

Python Programming for Arduino

ISBN: 978-1-78328-593-8   Paperback: 400 pages

Develop practical Internet of Things prototypes and applications with Arduino and Python

1. Transform your hardware ideas into real-world applications using Arduino and Python.

2. Design and develop hardware prototypes, interactive user interfaces, and cloud-connected applications for your projects.

3. Explore and expand examples to enrich your connected device's applications with this step-by-step guide.

Please check **www.PacktPub.com** for information on our titles